Employee Resourcing in the Construction Industry

T0259052

Employee Resourcing in the Construction Industry

Strategic considerations and operational practice

Ani Raidén, Andrew Dainty and Richard Neale

Routledge
Taylor & Francis Group

LONDON AND NEW YORK

First published 2009
by Taylor & Francis

This edition published 2014
by Routledge
2 Park Square, Milton Park, Abingdon, Oxon OX14 4RN

Simultaneously published in the USA and Canada
by Routledge
711 Third Avenue, New York, NY 10017, USA

*Routledge is an imprint of the Taylor & Francis Group,
an informa business*

First issued in paperback 2016

Typeset in Sabon by
RefineCatch Limited, Bungay, Suffolk

British Library Cataloguing in Publication Data
A catalogue record for this book is available from the British Library

Library of Congress Cataloging-in-Publication Data
Raiden, Ani B.
    Employee resourcing in the construction industry : strategic
considerations and operational practice/Ani B. Raiden, Andrew R. W.
Dainty and Richard H. Neale.
        p. cm.—(Spon research, 1940–7653)
Includes bibliographical references and index
    1. Construction industry—Personnel management.   I. Dainty,
Andrew.   II. Neale, R. H.   III. Title.
    HD9715.A2R34 2009
    624.068′3—dc22

                                                        2008033655

ISBN 13: 978–0–415–37163–6 (hbk)
ISBN 13: 978–1–138–96871–4 (pbk)

Contents

Figures and tables

Figures

Tables

Foreword

Researchers in the field of strategic human resource management are often attracted to an analysis of the impact of employee resourcing strategies and practice on organisational performance. As a consequence a variety of theoretical frameworks have evolved that relate organisational 'success' to the core functions of recruiting, developing and retaining key workers. The application of progressive employee resourcing strategy practices is seen as necessary to preserve a superior workforce within global industries.

This book focuses on the construction industry and the challenges that it faces in adopting a more strategically orientated approach to employee resourcing. The issues are clearly explored throughout the text, with a persuasive case for a more enlightened appreciation of the benefits of adopting a strategic approach to managing people being made. Difficulties in the implementation of progressive HRM practices and procedures in terms of employee resourcing are acknowledged and are in part a reflection of the complexity and disparate nature of the construction industry. Amongst the key challenges identified throughout the research is the discontinuity between the aspirations of senior managers within the industry for a more enlightened and positive appreciation of the value of a strategic focus to HRM and those of front line managers. Improvements in communication and opportunities for training and development of such managers are identified as being an important step to improve the operational effectiveness of such managers in adopting and maintaining a more strategically orientated understanding of the value of effective HRM practice.

The Strategic Employee Resourcing Framework (SERF) presented within the book does not set out to be a prescriptive tool for adoption within the industry, but a mechanism through which organisations might reflect on the efficacy of their own practices. As such this research monograph will appeal to a wide audience of both theorists and practitioners both within and outside of the global construction industry and can serve as a conduit for discussion and reflection on current industry practice.

Harry Barton
Head of Human Resource Management
Nottingham Business School

Preface

The special characteristics of the construction industry provide a stimulating but challenging environment for the development of progressive human resources management (HRM). The *structural context* within which HRM has to operate within the industry is project-based. Generally, construction works are large, complex, located and constructed on the site where they are required rather than in a factory. They are usually designed for specific purpose and location and so are unique, and often with an international dimension in their locations or personnel or both. These projects demand a very wide range of skills, which may differ quite radically from project to project, and the skills required also change during the projects – for example from groundworks (excavation, drainage and roadworks) to structural concrete to finely finished cladding and through to elegant interior work. In addition to the structural context there is a very strong *cultural context*. Hard physical work, often in unpleasant conditions, with a work force conditioned appropriately, generates its own special culture. Working on such substantial and socially useful projects can be very personally motivating, but the nomadic lifestyle and work culture can cause serious problems with an individual's work–life balance and general well-being. Thus, there are both the positive and negative aspects of working in the construction industry.

The demands on those who manage projects within this context are very dynamic and technically and spatially demanding. They have to mobilise disparate groups of itinerant workers, many of whom have little affinity to either the principal contractor or to the projects on which they work. Indeed, the fractured workplace structure has been shown to render the industry one of the most demanding environments within which to manage people. Add to this the need to ensure management succession and the need to meet the longer term developmental needs of both the organisation and their staff and the scale of the challenges explored within this book are clear.

This research monograph discusses these issues in-depth and proposes a framework of HRM principles and practices for addressing them. It does not attempt to achieve this through management prescription, but by proposing fresh ways of rethinking and connecting the issues at stake in the construction resourcing process. We have worked as a team on practical, industry-based research for many years, blending our respective experience in the management of human resources, strategic management, project management and production management in the specific context construction industry. The empirical insights gained through our collaborative work together have been used in order to convey the scale of the people management and resourcing challenge in the construction industry, but also to offer some novel solutions and fresh research trajectories for the industry to address in the future. We hope that readers will find this a stimulating mix of relevant theoretical and practical perspectives and that this will inform future developments in improving the way in which people are managed and developed within the construction sector.

Ani Raidén, Andrew Dainty and Richard Neale
Nottingham, Loughborough and Cardiff, UK

About the authors

Ani Raidén is Senior Lecturer in Human Resource Management at Nottingham Business School. She studied for her PhD (on which this book is based) at Loughborough University, Department of Civil and Building Engineering. Her research on employee resourcing and human resource development in the construction industry has been published in both construction management and HRM journals and conferences.

Andrew Dainty is Professor of Construction Sociology at Loughborough University's Department of Civil and Building Engineering. A renowned researcher in the field of human resource management and organisational behaviour in the construction industry, he has published widely in both academic and industry journals.

Richard Neale is a civil engineer who worked for consulting civil engineers and construction contractors before becoming a lecturer at Loughborough University and then a Professor of Construction Management at the University of Glamorgan. Richard retired in 2008 and is now Professor Emeritus.

1 Introduction

This book is based on an extensive programme of research undertaken in collaboration with seven major construction contractors in the UK. The aim was to examine the ways in which leading construction companies managed the complex but vital processes inherent in employee resourcing. As a research monograph, it describes some of the major insights derived from this study and explains possible new directions for both practitioners and researchers interested in how to better manage people in dynamic project-based organisations. The main objectives of the book are to provide the readers with the following:

- a comprehensive review of current HRM principles, procedures and practices, and their potential application to construction companies;
- a framework for implementing a strategic approach to HRM.

This book is not so much about developing new ways of working, but focuses on drawing out and synthesising promising practices used by different organisations grappling with the complexities of balancing their short- and longer-term employee resource needs with the personal development goals of their employees. It attempts, therefore, to bring together the elements of effective people management evident in company practices, but through doing this it seeks to develop new ways of overcoming the traditional constraints which have adversely affected resourcing practice in the past. The suggestions made are not designed to be 'best practice' prescriptions, but possible developments which organisations can mould and reshape to reflect their own unique ambitions and needs. The early part of this book demonstrates that such strategies, supported with appropriate information technology, can bring substantial benefits to the satisfaction, motivation and well-being of employees as well as significant savings on costs, and thus, return on investment for the business. The hope is that, by the end of this book, readers will see that realising such benefits is also possible within construction organisations.

The book addresses both strategic considerations and operational aspects of the people management function as applied to large contracting organisations. The text is organised into six chapters. After the introduction (Chapter 1), two chapters examine strategic Human Resource Management (HRM) and the operational aspects of the key HRM functions: employment relations, learning and development and employee resourcing. Cross-cutting themes such as work–life balance, technology and flexibility are discussed in relation to the demands of the contemporary construction firm. Chapter 4 presents a critical discussion on empirical evidence gathered from seven large construction firms. This culminates into the identification of five key themes: teams, human resource planning, performance management, employee involvement, and learning and development. These are combined to form a Strategic Employee Resourcing Framework (SERF) in Chapter 5, before conclusions are drawn and future directions for research and practice are articulated in Chapter 6.

1.1 The construction industry people management challenge

Although texts abound on how to manage people at both strategic/operational levels, the extent to which effective HR practices ensure improved organisational performance is far from uncontested ground. Within the context of construction, the complexity and dynamism of the industry's project-based nature render the applicability of many central tenets of strategic HRM tenuous (Loosemore *et al.*, 2003). It is important, therefore, to firstly examine the nature and context of the sector as a backcloth to exploring how these operational challenges can be overcome by more rigorous practices in the future.

The construction industry is one of the largest and most complex sectors. In fact, it is a very difficult sector to pin down and define, with some arguing that it is, in fact, an amalgamation of several sub-industries (see Ive and Gruneberg, 2000). In a recent report which explored the social and economic value of the industry, Pearce (2003) suggested that there was a narrow and a broad definition of construction. The narrow definition, based around the areas covered by the Standard Industrial Classification (SIC 45), excludes many aspects of what could be considered 'construction activity', such as design and engineering services. Taking a broader definition means that construction accounts for around 10% of gross domestic product (see Dainty *et al.*, 2007). Indeed, if SIC categories 45 and 74.2, are considered, then employment can be estimated at around 2.4 million, with a likely increase to 2.8 million by 2012 (Construction Skills, 2008). Thus, the industry is one of the country's major employers.

One of the key characteristics of construction work is that the demand for its services and products is highly variable, and particularly susceptible to economic fluctuations. This along with the industry's project-based structure has fundamental effects on the way in which people are employed and managed within the sector. Cherns and Bryan (1984) developed the term 'temporary multiple organisation' to describe project organisation in this regard. Projects typically comprise a complicated set of temporary inter-organisational relationships, which are themselves governed by project-defined interactions (see Bresnen *et al.*, 2004).

Raidén and Dainty (2006) recognise that, since a key feature of the industry's output is that the finished product is largely non-transportable and must therefore be assembled at a point of use, construction organisations must set up their temporary organisational structures at dispersed geographical locations, frequently at a distance from central management. Thus, the project team forms the focus of working life in construction, operating with a significant and necessary degree of independence. In addition, the changing requirements of construction work necessitate the formation of bespoke teams each time a new project is awarded. However, the time available between contract award and the mobilisation of the project is usually extremely limited (Druker *et al.*, 1996: 407). This renders planning for such deployment difficult.

The fluidity that these characteristics create leads to an unstable employment which necessitates a highly mobile and itinerant labour force. Workers move from project to project, often working as sub-contracted or self-employed workers. This has serious implications for the governance of projects and places considerable pressures on those leading the production effort on site. As Dainty *et al.* (2007) explain, the reduction in employed labour since the 1970s has left most of the firms at the head of the production effort doing little more than managing the construction process. Hence, it is increasingly important that their site management teams have the right blend of skills and competences in order to ensure successful project outcomes. However, construction companies must balance project requirements with competing organisational and individual employee expectations, priorities and needs. It is the industry's inability to manage these competing demands effectively which has caused many of the enduring problems for the sector. Finding a solution for overcoming these complexities is the central challenge addressed by this book.

The research presented within this text should be viewed within the context of the prevailing economic conditions, and particularly the tight labour market, which were affecting the industry at the time that the work was conducted. During periods of sustained growth staffing issues are often felt more acutely in terms of demand for new recruits. During a recession, attention inevitably turns to carefully selecting staff who are able to maintain competitive advantage despite reduced workload or tighter financial constraints. Hence, the importance of developing effective strategies for recruiting, developing and retaining high performing staff is not diminished in times of economic downturn; staff retention, development and reward remain fundamental to the competitiveness of construction firms. In times of growth, with staff turnover figures for construction organisations running at 21.7% (CIPD, 2007), the imperative to address reactive HR decision-making is simply rendered a more visible priority.

1.2 The research base of this book

As explained in the introduction, this book is based on an extensive programme of applied research conducted with seven major construction organisations. The aims of the research were as follows:

- to produce a structured and comprehensive explanation of current
  employee resourcing practices within large construction contractors;[1]
- build a framework for the development of strategically aligned management practices for effective staff deployment.

In light of this, initially, the research objectives were to establish key organisational HRM strategies, policies, practices, organisational priorities and
project requirements in relation to employee resourcing; and model their
current resourcing decision-making processes. Simultaneously, the research
sought to establish employees' personal and career needs and preferences
in relation to their deployment. These two data sets were then compared
and contrasted in order to establish the compatibility and conflicts between
managerial (organisational and project) and employee deployment objectives. Finally, a framework was developed that explains the employee
resourcing processes reflecting leading-edge practice within the industry's
larger employers.

The research methodology

Investigation of organisational processes and priorities, project requirements and employee needs and preferences is a complex task, requiring a
systematic approach to data collection and analysis if meaningful results are
to be achieved. An overall interpretative framework was used; that is, the
philosophy of the researchers was to try to investigate the processes, practices and attitudes from the point of view of the companies and individuals
involved, rather than to work from the researcher's perspective. A case study
methodology was adopted as an approach to the investigation, because this
would yield in-depth information on the reality of the issues. This included
one in-depth 'primary case' and six 'supplementary cases'. The study organisations were carefully selected to be broadly representative of leading large
UK-based contractors: similar in size, number of employees and turnover.
Project-specific case study data were also collected within the primary case
study organisation.

Case study organisations

The primary case study consisted of a national contractor, with an annual
turnover in excess of £550 million during the research period. In the three
years of research, the company experienced rapid growth in turnover, business development and share value. The company activities were distributed

among water engineering (34%), rail and highway development (20%), commercial development (19%), education and health (11%) and, to a lesser extent, housing (8%), industrial works (6%) and other smaller projects, such as interior and refurbishment business (2%). Clients within these sectors included both public and private developers.

The company employed approximately 2,100 staff. These were managed in three regional divisions: North, Midlands and South. The regions contained independent operational divisions and smaller departments, each of which served a distinct sector of the market. These divisions and departments operated as individual profit centres. Typical to the industry, the organisation had recently merged with a similar contractor. Chapter 4 discusses the structure and operating mechanisms of the company in detail.

The supplementary cases selected were similar to the in-depth case study organisation in size, number of employees and turnover. Table 1.1 lists the type of participating organisations, their main operating sectors and annual turnover at the time of the research interviews. 'National contractor' refers to an organisation with its main base (headquarters) in the UK. This distinguishes case studies A–E from the 'European contractor' whose headquarters are located in Europe. All of these organisations have a strong international profile.

Research methods

The research commenced with an in-depth investigation into the existing resourcing process. Interviews were held with HRM staff and senior managers within the primary case study organisation in order to establish the organisational policy and practice in terms of resourcing and staff development. Four project case studies were then undertaken in order to explore the efficacy of historical deployment and project allocation decisions. The composition of teams selected for a range of projects of differing size and complexity (multi-site, PFI, design and build and traditionally procured) were explored and interviews held with both the line managers responsible for the resourcing decisions and with the team members themselves. This

Table 1.1 Supplementary case study profile

Case study	Sector	Turnover (£'000s)
A National contractor	Building	1,074.000
B National contractor	Civil engineering	487.238
C National contractor	Civil engineering	1,680.000
D National contractor	Building	1,382.200
E National contractor	Civil engineering	1,335.900
F European contractor, UK branch	Building	706.000

was followed by a series of in-depth interviews with other project-based employees in order to establish their personal priorities in terms of career aims and development, project allocation and their wider life-cycle priorities. The interview data were combined using a qualitative data analysis software package, which was also used to construct a model of the existing resourcing process from which strengths and weaknesses could be established. A range of secondary data were collected and analysed in order to inform wider understanding of the existing approaches to the deployment process. These included an analytic hierarchy method questionnaire, which was used to refine a list of factors important to employees when making project deployment decisions; a factor verification questionnaire that confirmed the interview findings within a wider sample; and an assessment of the senior managers' management style; together with organisational documentation.

Interviews

Altogether 50 respondents were interviewed as part of the primary case study. This included divisional directors (4), senior operational managers (7), specialist HRM practitioners (4) and employees at various levels of the organisation (35). Contact was established with some of these respondents more than once. Respondents within the supplementary cases were human resource and operational senior managers responsible for project allocation decision-making (9).

A bespoke interview schedule was developed for establishing the following:

- the organisation's HRM strategy, policy and practice;
- the organisational and project requirements;
- the individual needs and preferences of employees.

The schedule provided a fairly loose structure for the discussions and helped to ensure all the topics were covered. However, the aim was not to restrict issues from emerging. Interviews within the supplementary cases concentrated on exploring a range of innovative approaches to the resourcing process. This widened the research perspective beyond the principal collaborating company and, together with literature, these data were used to develop a showcase of effective practice. Although no single company had developed a completely successful integrated approach to managing the resourcing process, a combination of ideas and perspectives from each organisation informed the understanding of the resourcing issues holistically.

All interview material was tape-recorded, transcribed verbatim and analysed using qualitative data analysis software. It allowed for the data to be analysed, for example, by correlating responses by respondents' job role. In addition to the demographic details of the respondents, a document analysis and three questionnaires were administered within the primary case study in order to triangulate the qualitative interview material.

Analytic hierarchy method

The analytic hierarchy method questionnaire (Saaty, 1980) asked employees to rank the importance of nine factors that potentially influence team deployment decision-making against each other:

- personal and/or professional development
- work location close to home/maintaining work–life balance
- training opportunities
- organisational division
- experience in working under different procurement systems or contract forms
- gaining broad and/or specialist experience
- project type (e.g., size, complexity, etc.)
- good team relationships
- promotional opportunities.

The results were analysed statistically, and then linked to the qualitative interview data.

Factor verification questionnaire

The factors identified as important to be taken into account in the employee resourcing process within the interview and analytic hierarchy method questionnaire data were verified via a self-administered postal questionnaire. This was carried out in order to do the following:

- confirm the factors that had been extracted truthfully;
- establish an estimated rank order for the variables included;
- allow respondents beyond the original interview sample to inform the research of any additional factors that had not previously been identified.

The questionnaire was designed as a follow-up on the primary case study organisation's annual staff satisfaction survey. The forms, together with self-addressed prepaid return envelopes, were delivered to the organisation's main research contact, who distributed them across the organisation. Departmental managers encouraged their staff to complete and return the forms. This approach adopted for the administration of the questionnaire helped to secure a good response rate. The results were analysed statistically.

Management style questionnaire

The Blake and Mouton's (1985) managerial grid technique was used to measure the senior management team's leadership style along two dimensions: concern for people and concern for production. The questionnaire form included six elements, which asked the respondents to select a statement that most closely matches their view on the following areas:

- *Decision-making*: I accept the decisions of others with indifference/I place high value on sound creative decisions that result in understanding and agreement.
- *Convictions*: I avoid taking sides by not revealing opinions, attitudes and ideas/I listen for and seek out ideas, opinions and attitudes different from my own. I have strong convictions but respond to ideas sounder than my own by changing my mind.
- *Conflict*: When conflict arises, I try to remain neutral/When conflict arises, I try to identify reasons for it and seek to resolve underlying causes.
- *Emotions* (temper): By remaining uninvolved, I rarely get stirred up/ When aroused, I contain myself even though my impatience is visible.
- *Humour*: My humour is seen as rather pointless/My humour fits the situation and gives perspective; I retain sense of humour even under pressure.
- *Effort*: I put out enough to get by/I exert vigorous effort and others to join in (Blake and Mouton, 1985).

The analysis was carried out by spreadsheet and results were plotted onto a grid.

Document analysis

The analysis of the organisational documentation involved the examination of all available and relevant printed company information, such as

- annual reports and mission statements;
- policies and procedures (including equal opportunities policies, new employee induction packs, general terms and conditions of employment, offers of employment and performance appraisal forms);
- project outcome records (customer project reviews);
- leaflets, brochures, pamphlets and booklets published for advertisement and/or PR purposes.

The summary accounts were coded in a similar fashion to the interview material, and so incorporated in the search and retrieval process.

Human resource information systems (HRIS's) in construction surveys

As a preliminary explanation of the use of HRISs in construction, in addition to the case study material, a short postal questionnaire survey was administered to 100 leading medium–large construction organisations (in 2001). The respondents were asked to state their use of information technology applications for human resource related functions; which HRIS application, if any, they used; the length of time the system had been in place; the functions for which the HRIS was used; and how satisfied they were with the system. The questions were designed to allow a comparative analysis with the annual Computers in Personnel Survey which is carried out by the (CIPD) and Institute for Employment Studies. This established survey has charted the increasing utilisation of IT across a variety of different sectors over recent years.

Following the initial returns of the questionnaire, a series of telephone interviews was conducted with a sample of the original informants. Respondents who had showed interest in the results of the survey and specified using HRIS for appraisal records and/or to deploy staff to projects were contacted and asked to elaborate on their answers to the questionnaire. This added qualitative data to complement the survey results such as details of user experiences and utilisation characteristics.

The questionnaire was administered again in 2007 to survey any changes in the use of HRISs in large construction organisations in the UK. Both in 2001 and in 2007 the results were analysed using MS Excel.

Data analysis

In terms of a unit of analysis, the research involved three parallel levels of analysis: the organisation, projects and individual employees. Figure 1.1 illustrates how these came together to form a comprehensive framework of analysis.

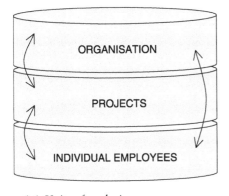

Figure 1.1 Units of analysis.

At the level of the organisation, the current strategic HRM strategy, policies and processes, employee resourcing activities and factors that specify the organisational priorities were analysed. Real-life examples are used to reveal how these practices played out in practice. These allow for an examination of the project requirements managerial respondents highlighted as important to take into account in the resourcing decision-making. The individual employee account provided their perceptions of the effectiveness of the organisational strategy, policy and practice. The respondents' personal needs and preferences were extracted and analysed in relation to the current practice and organisational project priorities and requirements.

Industrial collaboration

This research was funded by the Engineering and Physical Sciences Research Council (EPSRC). In kind support was also provided by the seven participating companies who allowed their managers and employees to take part in the research process. The principal research partner in particular provided extensive support for the work in terms of 50 informants and almost unlimited access to organisational and project data. Their interest in the research stemmed from a need to improve the recruitment and retention of newcomers into the organisation due to recent organisational growth and rapidly increasing workloads.

Care has been taken to involve the industrial partners at every stage of the research through the production of regular reports and project review meetings. These have ensured the practical utility of the outputs as well as the timely realisation of the project's outcomes (see Chapter 6).

1.3 Structure of this book

The book largely follows the thought process discussed in the Introduction. A foundation of theory and contemporary thinking is provided in Chapter 2, which charts the evolution of strategic HRM. This discussion is elaborated in Chapter 3, where the main components of strategic HRM are explored in detail. Chapter 4 critically reviews current employee resourcing practices. This leads into the SERF (Chapter 5). Chapter 6 concludes the book with a summary of the achievements of the research, together with a discussion of its limitations and suggestions for future directions in both research and practice around the resourcing issues discussed within the book.

Note

1   Large contractor is defined as employing 600+ employees (DTI, 2003).

2 The evolution of contemporary strategic HRM

The first chapter set the context for the book and introduced the general background to managing people in the construction industry. This chapter investigates the theoretical foundations of strategic Human Resource Management (HRM). It begins by exploring the origins and various models of strategic HRM. Theories in organisational behaviour, a major contributory field to HRM, are outlined. The differences between industrial relations, the personnel management paradigm and strategic HRM are discussed in relation to the emerging realisation of the importance of effective people management as the key to competitive advantage. This chapter also examines the current debates and cross-cutting themes in the strategic HRM literature, such as high-performance work.

2.1 Defining strategic HRM

Like many concepts in management-related subjects, strategic HRM is not easy to define. The specific difficulties in understanding what strategic HRM derive from the two quite distinct ways it is used. Firstly, many textbooks employ the term generically to describe a range of management activities that relate to managing people. This generic term often encompasses organisational/employee resourcing related objectives such as staffing, performance management, HR administration and change management (see Taylor, 2005; Torrington *et al.*, 2005). These are discussed in Chapter 3. The second meaning commonly associated with strategic HRM specifies a particular approach to management of people. This view specifies nuances in the ways in which people management activities are undertaken, for example whether the focus is on the workforce specifically, or resources more broadly. In this book, strategic HRM is considered as a useful route to good people-management practices, particularly when it is not tied to any prescriptive dimensions.

Leading textbooks in HRM take slightly different angles to explain the historical development of the strategic HRM concept. For example, Bratton and Gold (2003) focus on the impact of product and labour markets, social movements and public policies on shaping current societal beliefs. Beardwell and Claydon (2007) discuss these largely in the context of developments in the USA. Torrington *et al.* (2005) developed a more organisational focus on management considering five themes:

1   Social justice – Early criticism of the free enterprise system and the hardship created by the exploitation of workers in factories helped create the first posts for personnel managers.
2   Humane bureaucracy – Move away from sole focus on welfare, added responsibilities for staffing, training and organisation design.
3   Negotiated consent – Bargaining and negotiations with Trades Unions.
4   Organisation – Integration into managerial activity, development of opportunities for personal growth.
5   HRM (now strategic HRM) – A range of management activities related to managing people vs a specific approach to people management.

This structure informs the journey from organisational behaviour to strategic HRM explained in the next section. This outlines the major approaches and theories in organisational behaviour that shape strategic HRM. Following this, industrial relations, personnel management and HRM are discussed before a consideration of *strategic* HRM through models, contemporary debates and its key components in Chapter 3. Clearly, such a summary is an overt simplification of what is a highly contextual issue. For example, bargaining and negotiations with Trades Unions have little influence on construction organisations (Druker, 2007). The purpose here is to provide an overview of the historical developments in different approaches to managing people.

Organisational behaviour

The study of organisational behaviour is concerned with, on the one hand, the behaviour of individuals in the organisation, and on the other, groups and how they form, perform, change and develop. Communication and perception, motivation, learning and personality form the key to understanding the ways in which one person interacts with another person (Buchanan and Huczynski, 2006). Central to understanding groups in organisations are issues of group formation, development, structure, control/cohesiveness and effectiveness (Belbin, 1993, Tuckman, 1965; Mullins, 2005). The origins of organisational behaviour can be traced back to the 1920s in the form of the human relations movement. Since then, numerous theories have emerged within the field to facilitate the understanding and analysis of individual and group behaviour within organisations. Table 2.1 presents some of the general theories in organisational behaviour. Tables 2.2 and 2.3 present theories related to the individual as the unit of analysis and Table 2.4 focuses on those theories that explain the dynamics specifically related to groups.

Table 2.1 Some general theories of organisational behaviour (after Buchanan and Huczynski, 1997, 2006; Huczynski and Buchanan, 2001; Mullins, 1996, 2002, 2005)

Approach	Theory	Time period	Characteristics
Scientific management		Frederick Taylor (1870–80)	Task fragmentation; One best way; Training for simple and fragmented tasks; Reward piece rate
Human relations approach	Hawthorne studies (1924–33)	George Elton Mayo (1920s)	Demonstrated the influence of social factors on workplace behaviour
	Socio-technical systems	Trist (1940s)	Studies on effects of changing technology – a sub-category of the industrial relations systems theory, provides a link between systems theory
	Technology approach	Walker and Guest (1950s), Sayles (1950s), Blauner (1960s), Woodward (1980s)	Emphasises the effects of varying technologies on organisation structure, work groups and individual performance and job satisfaction – a sub-category of systems theory
	Decision-making approach	Barnard (1940s), Simon (1970s)	Focus on managerial decision-making and how organisations process and use information in decision-making
Contingency approach	Attempts to analyse organisation structure in terms of relationships among its components and the environment with emphasis on flexibility		
		Burns and Stalker (1960s)	Mechanistic/organic system
		Lawrence and Lorch (1960s)	Differentiation/integration
	'Peter principle'	Peter and Hull (1970s)	Study of occupational incompetence and organisation hierarchy
	Shamrock organisation	Handy (1980s)	Study of flexible organisation design
	'Parkinson's law'	Parkinson (1980s)	'Rising pyramid' – work expands as to fill the time available for its completion

Table 2.2 Communication, perception and motivation theories related to the individual (after Buchanan and Huczynski, 1997, 2006; Huczynski and Buchanan, 2001; Mullins, 1996, 2002, 2005)

Approach	Theory	Time period	Characteristics
Communication and perception	'Proximity' and 'similarity' principles	Max Wertheimer (1923)	Group together or classify stimuli – in person perception these are applied in assuming people are similar on the basis of certain factors (i.e. nationality)
	Impression management	Erving Coffman (1959)	The process whereby people seek to control the image others have of them
	Halo effect	Edward Thorndyke (1920)	On meeting a stranger, people 'size them up'/make judgements about the kind of person they are
	Stereotyping	Walter Lippmann (1922)	Grouping people together who seem to us to share similar characteristics
	Attribution theory	Fritz Heider and Howard Kelley (1950–60)	Attaching or attributing causes or reasons to the actions and events we see
Motivation Goals	Maslow's need hierarchy	Abraham Maslow (1943)	We have 8 (9) needs, which each need satisfying in an order, the goal being self-actualisation and transcendence
Cognitive decision process	Expectancy theory	Edward E. Tolman (1930s)	Behaviour depends on the outcomes that an individual values
Social process/ classical		Victor H. Vroom (1964)	1st systematic formulation of expectancy theory – measuring motivation on valence and subjective probability
		Henry Ford	Mass production
	Two factor theory	Frederick Hertzberg (1950s)	Characteristics of work influence job satisfaction and dissatisfaction

Table 2.3 Learning and personality theories related to the individual (after Buchanan and Huczynski, 1997, 2006; Huczynski and Buchanan, 2001; Mullins, 1996, 2002, 2005)

Approach	Theory	Time period	Characteristics
Learning			
Behaviourism		John B. Watson (1913)	Reward is more effective than punishment in changing behaviour – learning is the development of associations between stimuli and responses through experience
	Pavlovian (or classical/ respondent) conditioning	Ivan Petrovich Pavlov (1850–60s)	Technique for associating an established response with a new stimulus
Cognitive approach	Skinnerian (or instrumental/operant) conditioning	Burrhus Frederic Skinner (1920–30s)	Technique for associating a response or a behaviour with its consequence
	Information-processing theories	Norbert Wiener (1947)	Draws from cybernetics – control of (system) performance through feedback
	Social learning theory	Albert Bandura (1977)	Learning through observation and copying others without any reward/ punishment
Personality			
Nomothetic approach	Extraversion– intraversion 'E' and neuroticism stability 'N' dimensions	Hans Jurgen Eysenck (1940s)	Based on genetics, a way of linking type, traits and behaviour
	Types A and B	Meyer Friedman and Ray Rosenman (1974)	Two extreme personality patterns or behaviour syndromes which help to explain differences in stress levels

Table 2.3 Continued

Approach	Theory	Time period	Characteristics
Idiographic approach	'Looking glass self'	Charles Horton Cooley (1880–90s)	Personality of an individual is the result of a process in which individuals learn to be the person they are
	Generalised other	George Herbert Mead (1930s)	What we understand other people expect of us in terms of attitudes, values, beliefs and behaviour
	Two-sided self and impact of accepting environment	Carl Ransom Rogers (1940s)	The core of human behaviour is a desire to realise one's potential fully, however right social environment (where one is treated with 'unconditional positive guard') is required
	A thematic apperception test (TAT)	Henry Alexander Murray (1938)	'Projective' assessment in which an individual is invited to project his/her own interests and preoccupations into accounts of pictures/stories
		David McClelland	Developed the TAT since Murray

Frederick Winslow Taylor developed an early understanding about manufacturing workflow and productivity in the 1880s and 1890s. His early theories of management were called 'scientific management'. Drawn from economics and engineering, this 'hard' approach segregated planning from doing and rewarded increases in piece rate output. The Human Relations Movement introduced 'softer' elements to management thought. Elton Mayo revolutionised our thinking through the Hawthorne studies. His relay assembly and bank wiring room experiments sought to identify the impact of different variables on productivity. Positive change was apparent, but for reasons unexpected. Informal social pattern of the workgroup was more important than physical conditions or financial incentives. Through industrialisation of the society and advances in technology 'socio-technological systems' and 'technology' approaches were developed. These emphasised the

Table 2.4 Group theories of organisational behaviour (after Buchanan and Huczynski, 1997, 2006; Huczynski and Buchanan, 2001; Mullins, 1996, 2002, 2005)

Approach	Theory	Time period	Characteristics
Group formation		Muzater Sherif (1949–53)	Experiments on informal group formation, their status structure
	Homan's theory of group formation	George Caspar Homans (1950s)	Any social system exists within a three-part environment: physical, cultural, technological. This external system influences behaviour causing an internal system to arise. Changes in either affect the other, they are interdependent.
	Linking pin model	Rensis Likert (1960s)	Structure of an organisation should be formed around effective work groups as group forces are important in influencing the behaviour of individual work groups and entire organisations
		Levitt (1970s)	Management should consider building organisations using small groups.
		Tom Peters (1980s)	'The modest-sized, task-orientated, semi-autonomous, mainly self-managing team should be the basic organisation building block'

effect of introducing technology to organisational (and staffing) structure, dynamics of work groups and job satisfaction.

From the 1960s, a great deal of attention was focused on the 'contingency approach'. Flexibility formed the central concern. Belief that 'there is no one right way to manage' applied to the analysis of organisational structures and cultures led to the development of theories of an organisation as an organic/ mechanistic system and integration vs differentiation. Handy's (1980) Shamrock organisation and Atkinson's (1984) flexible firm illustrated organisational designs for flexible employment strategies that are still valid (see Chapter 3 for more detailed discussion on this).

Much of the early work on communication and perception theories has influenced our understanding and development of fair decision-making (Table 2.2). For example, the halo effect, stereotyping and attribution theories are central to recruitment and selection. Similarly, motivation theories are considered in performance management and reward in particular. HRM texts provide an in-depth discussion on the contribution and criticisms of these theories applied to organisational HRM practice (Bratton and Gold, 2003), leadership (Torrington *et al.*, 2005), learning and development (Grugulis, 2006), reward (Leopold *et al.*, 2005) and performance management (Beardwell and Claydon, 2007).

The organisational behaviour theories in learning draw on basic principles of human psychology. Many of these ideas form the foundations of learning and development, what has grown into a more sophisticated field on its own right. Much of this develops on the principles of social learning theory and draws on other areas such as motivation. Learning is seen as a reward in itself. Chapter 3 discusses this in detail.

Another area of psychology that has contributed to the development of organisational behaviour and HRM is 'personality'. This has become an explicit consideration in recruitment and selection where HRM professionals seek to increase the predictive validity and fairness of the decision-making process. Other uses include personal development, leadership coaching and group dynamics. The nomothetic approach, which classifies different types of personalities, has been particularly influential. From extraversion–introversion scales were developed psychometric testing tools such as the Myerss Briggs Type Indicator. This questionnaire-based assessment is used widely to find out individual's preferences in four areas:

- extraversion vs introversion
- sensing vs intuition
- thinking vs feeling
- judging vs perceiving.

Understanding your own and others' broad preferences can help in cohesive groupwork. Many early theories in group formation focused on convincing management that small groups, or teams, were the most effective way of organising people (Table 2.4). In light of this evidence, it is not surprising that groups, or teamwork, are heavily integrated into organisational life, in general, and in particular, the construction industry. As discussed in the introduction (Chapter 1) autonomous groups are central to construction operations. Chapter 5 provides further insight into this important area.

As the study of organisational behaviour focuses on understanding, predicting and controlling human behaviour and the factors, which influence the performance of people as members of an organisation, a close relation with management theory and practice is evident. The application of the theories of organisational behaviour to the context of management has led to the development of a number of narrower disciplines, such as industrial relations, personnel, HRM and strategic HRM. An insight into the developments within these fields is crucial to understanding the current strategic HRM approach to the employee resourcing function. Figure 2.1 shows the relative approximate chronological development of these concepts and theories. These movements are further elaborated hereafter.

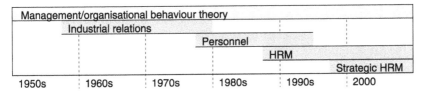

Figure 2.1 From organisational behaviour to strategic HRM.

Again, this illustration provides a deliberately overt simplification of a complex field of theoretical development and practical application. The central concepts of the different approaches hold relevant today. For example, larger organisations in particular have strong industrial relations systems in place, although some argue that these issues should be managed under the umbrella of strategic HRM which incorporates employee relations as one of the key components of the concept (see Chapter 3). The purpose here is to summarise some of the key historical developments that are useful in understanding the background to strategic HRM.

Industrial relations

Industrial relations laid the foundation for effective people management since it provided the overall framework for the design and implementation of strategies, policies and processes. The early theories in industrial relations emerged in the late 1950s, dominating much of the people management practice and research through to the development of the concept 'personnel' in the 1970s. The following provides an overview of the key theories in industrial relations, such as Dunlop's systems theory, the unitary theory and its neo-unitary variant, conflict-pluralist theory, social action theory, Marxist theory, and their different levels of individualism and collectivism.

Systems and social action theories

Dunlop's simple model of an industrial relations system 'presents a general theory of industrial relations and provides tools for analysis to interpret and gain understanding of the widest possible range of industrial relations facts and practice. For Dunlop, an industrial relations system is not part of a society's economic system, but a separate and distinctive subsystem of its own, partially overlapping the economic and political decision-making systems with which it interacts' (Farnham and Pimlot, 1990: 10–11).

In this model, an industrial relations system comprises three factors: the actors, the contexts and the ideology. The actors comprise a hierarchy of managers and their representatives, a hierarchy of non-managerial employees and their representatives, and specialised third party governmental and private agencies. These actors interact with the three environmental contexts:

the technological characteristics of the work place and work community, the market or budgetary constraints, and the locus and distribution of power in the larger society outside the IR system (Dunlop, 1958). The ideology held by the actors binds the system together.

Although Dunlop's 'systems theory' has its critics, and has been refined and developed on numerous occasions, it has never been changed radically. Rather, it has influenced British and European IR as the major American contribution in the field (Farnham and Pimlot, 1990: 10). The relevance of the systems theory to the modern strategic HRM lies in its focus on industrial relations 'systems' as the institutional means by which the rules of employment are established and administered.

Social action theory compliments Dunlop's systems theory. The difference between social action and systems theory is that 'the action theory assumes an existing system where action occurs but can not explain the nature of the system, while the systems approach is unable to explain satisfactorily why particular actors act as they do' (Farnham and Pimlot, 1990: 10).

Thus, these theories are best used in combination. Action theory provides an analytical framework for assessing the factors that influence an actor's (e.g. manager's or an employee's) behaviour and hence can help explain why such behaviour occurs. The systems approach seeks to define and develop an understanding of the system and related processes within which the actor's behaviour occurs. Accordingly, the combination provides a holistic framework for analysing both the organisational system(s) as well as the behaviour of the actors within the system.

Marxist theory

In contrast to the systems or social action theories, Marxist theory highlights the class-influenced nature of the employment contract and the continuous struggle between those representing capital and labour (Torrington and Hall, 1991: 4–5). Although in its basic form it is somewhat outdated, Marxist theory can contribute to the consideration of the compatibility and conflicts between the needs of the employer and employee, which lie at the heart of strategic HRM. Indeed, in this Marxist theory has influenced the development of the modern 'employment relations' component of strategic HRM. This is discussed in Chapter 3.

Classical unitary, neo-unitary, industrial conflict and pluralist theories

More recently developed theories of industrial relations include the classical unitary and neo-unitary theories, together with the industrial conflict and pluralist theories. The classical unitary theory emphasises a stable structure and co-operative nature of work and work relations. The neo-unitary theory, on the other hand, being more sophisticated than the classical unitary theory, aims to integrate employees as individuals into the companies in which they work (Coupar and Stevens, 1998: 146). Its orientation is distinctly market centred, managerialist and individualist. Employers embracing this frame of reference have expectations of employee loyalty, customer satisfaction and product security in an increasingly competitive market place by gaining employee commitment to quality production, customer needs and job flexibility. The neo-unitary approach to managing people includes creating a sense of common purpose and shared corporate culture. It emphasises the primacy of customer service, setting explicit targets for employees, investing heavily in training and management development and providing employment security for their workers. This is achieved by using techniques, such as performance-related pay, profit sharing, harmonisation of terms and conditions and employee involvement to facilitate commitment, quality and flexibility (Storey, 1992).

The industrial conflict and pluralist theories focus on conflict identification and problem solving. They are concepts related to post-capitalism, within which society is viewed as an open system where political, social and economic power is dispersed (Farnham and Pimlot, 1990: 47–48). Moreover, since these approaches view society as comprising of a variety of individuals and groups, each holding divergent values and interests, of central importance to them is the accommodation of these different values, pressures and competing interests within organisations (Torrington and Hall, 1991: 8). The industrial conflict and pluralist theories are arguably the dominant theoretical approaches to industrial relations in Britain (Farnham and Pimlot, 1990, 1995).

Over time, as the theories of industrial relations developed towards facilitating employee welfare, commitment and quality, 'personnel' management practices began to gain importance within people management. The boundaries between industrial relations and personnel started to haze. In many organisations industrial relations has gradually become part of the personnel function orientating its focus on collective bargaining and negotiations. Indeed, 'employment relations' is now viewed as a central component of HRM, even though the most recent generation of HR managers are generally unaccustomed to dealing with disputes at work (Marchington and Wilkinson 2005: 265). An appreciation of the industrial relations theory is crucial to understanding the concepts of personnel, HRM and strategic HRM, since the early ideas on industrial relations underpin these modern approaches to people management. Chapter 3 discusses 'employment relations' further.

Personnel and HRM

The personnel management paradigm dominated the field during late 1970s and 1980s. Torrington and Hall (1991: 15) describe the function as being *workforce-centred* and directed mainly at the organisation's employees. They explain the role of personnel specialists to include the recruitment and training of employees, arranging for them to be paid, explaining management's expectations and justifying their actions to the employees, satisfying the non-managerial employees' work-related needs, dealing with their problems and seeking to modify management action that could produce unwelcome employee response. Key features of this style are focus on procedures and control, administration of employment contracts and job grades and collective bargaining with little strategic involvement (Sparrow and Marchington, 1998: 10). Although unquestionably a management function, personnel never totally identified with management interests; rather it focused on understanding and articulating the aspirations and views of the workforce. This paradoxical arrangement resulted in ineffective mediation of the needs of the organisation and those of the employees, since the personnel specialists' authority to influence change relied on implementation of personnel policy and procedures at operational level (Legge, 1989; Storey, 1992).

In the late 1980s, the concept of HRM made its way to Britain from the US bringing with it the strategic version, strategic HRM, less than a decade later. HRM shifted the focus from personnel administration towards training and development, organisational culture and performance-related reward mechanisms (Sparrow and Marchington, 1998: 10). In other words, people were viewed as a source of competitive advantage and as more valuable than other organisational resources. This was achieved through emphasis on planning, monitoring and control, rather than implementation of personnel/ HRM policy and procedures or mediation. Torrington and Hall (1991: 15– 16) view the function as being *resource-centred*, directed mainly at fulfilling the organisation's needs for human resources to be provided and deployed. The emphasis on human resource, rather than employees, had two major implications. Firstly, the HRM paradigm is concerned with the management and development of the management team (Storey, 1992). Secondly, it encourages employment flexibility through non-standard forms of employment, such as part-time work, self-employment and sub-contracting (Emmott and Hutchinson, 1998). This initiated changes to the way work is organised within organisations drawing attention to the tensions between the welfare-oriented industrial relations/personnel framework and business-driven need for human resource development. Strategic HRM moved this a step further with a focus on organisational redesign, broad set of competencies, human resource planning and tying various people management practices, initiatives and processes together (Mabey and Salaman, 1995).

Strategic HRM

Central to the concept of *strategic* HRM lies the argument that the effectiveness of an organisation largely depends on the efficient use of human resources. It comprises a set of practices designed to maximise organisational integration, employee commitment, flexibility and quality of work (Guest, 1987: 503). The elements and values of the concept are not new, but their combination and power to influence organisational change is novel (Mabey *et al.*, 1998: 36). It differs from HRM in its emphasis for relationships between structures and strategy and the environment external to an organisation (Tichy *et al.*, 1982; Fombrun *et al.*, 1984; Boxall, 1992). Devanna *et al.*'s (1984) matching model of strategic HRM is one of the early illustrations of this. The matching model (see Figure 2.2) clearly illustrates the strategic HRM emphasis on the relationships between human resources, organisational structures and strategy and the environment external to an organisation. It focuses on the integration of HRM policies with the organisational mission, strategy and structures as well as the (external) political, cultural and economic forces that influence the way an organisation is managed (Sisson, 1990). The model is useful in its plain and clear-cut view of the organisation in the context of the wider environment, however, this simplicity also presents a drawback in that it lacks sufficient detail for analytical purposes (Boxall, 1992). It is essentially a unitarist analysis of the HRM function in that people management is 'read-off' from the wider business objectives of the organisation (Marchington and Wilkinson, 2005: 5).

An alternative view to the matching model is the map of the HRM territory by Beer *et al.* (1984). This map of the HRM territory, or the Harvard model as it is better known, is probably one of the most influential illustrations of strategic HRM in that it embodies the environmental influences and factors internal to an organisation together with analytical components. The model provides an open-systems model for strategic HRM, which shows the situational factors' and stakeholder interests' influence on human resource policy, the human resource policies' influences on the

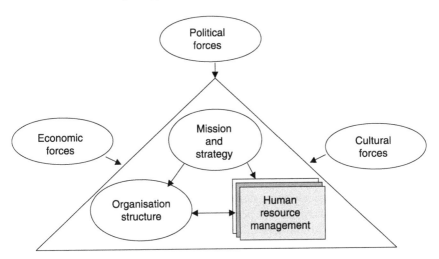

Figure 2.2 Matching model of strategic HRM (developed after Devanna *et al.*, 1984).

human resource outcomes and what the long-term consequences of operating a strategic HRM approach to people management may be (Beer *et al.*, 1984). The model also recognises that long-term consequences shape the situational factors and stakeholder interests, thereby providing a cyclical representation of the strategic HRM decisions, the business environment and an organisation's performance (Huczynski and Buchanan, 2001: 678). This link between the strategic HRM decisions, the business environment and an organisation's performance is crucially important. It adds the 'performance' factor into the model suggested by Devanna *et al.* (1984) earlier thereby providing a more analytically satisfactory representation of strategic HRM (Boxall, 1992). This incorporates a focus for the potential outcomes/ achievements an organisation might expect as a result of an integrated strategic HRM approach.

Although the Harvard model has gained much attention as one of the most influential illustrations of strategic HRM, it has been critiqued for its limitations to explain how the four policy areas (employee influence, human resource flow, reward systems and work systems) are influenced by the situational factors and/or stakeholder interests and how they might affect the human resource outcomes (Loosemore *et al.*, 2003: 40). The model does not explain *how* HRM should be considered as a strategic function. The assumed dominant direction of influence from situational and stakeholder interests oversimplifies what is, in reality, a complex and fragmented process whereby employers attempt to make policy choices in a structured way (Marchington and Wilkinson, 2005: 5).

Drawing on the prescriptive and analytical qualities of the Harvard model, Guest (1987) and Hendry and Pettigrew (1990) among others have developed British approaches to strategic HRM. Guest's contribution is realised in a set of propositions he developed: strategic integration, high commitment, high quality and flexibility, which he considered to be amenable to testing (Guest, 1987; Beardwell and Holden, 1997). The main contribution of Hendry and Pettigrew's strategic change and HRM, or Warwick model, is in that it incorporates culture and business outputs into the framework. The model reflects European traditions and management styles with an emphasis on a full range of tasks and skills that define HR as a strategic function. This compensates for the Harvard model's inability to explain the *how* in relationships between the different components of the HRM system.

The four HR elements in the Warwick model focus on the following:

- the use of planning;
- a coherent approach to the design and management of HR systems;
- matching HR activities and policies to business strategy;
- seeing people of the organisation as a 'strategic resource' for achieving competitive advantage (Hendry and Pettigrew, 1990).

The fifth element incorporates external factors, the environment, into the model. This completes the model as an analytical framework for assessing the impact of change on the strategic HRM function. Each element of the model reflects a particular context within which an organisation operates. This allows for a comprehensive view of the organisational factors and contexts to be established with the view of achieving competitive advantage through the effective deployment of people.

A more recent model, the integration of HRM systems by Sparrow and Marchington, (1998: 86) represents the complex relationships of the distinct but interdependent strategic HRM functions and processes in a star shaped design (Figure 2.3).

According to Sparrow and Marchington (1998: 86), the employee resourcing systems include role definition, resource planning, recruitment and selection, performance management and release from the organisation. The reward systems' focus is on monetary as well as non-monetary remuneration and benefits. The training and development systems provide suitable conditions for learning and the employment relations comprise of individual and collective communications systems and collective representation. The management of the HR function include team integration, organisational performance monitoring and management of the HR function. Each of these elements is discussed in more detail under *components of strategic HRM* in Chapter 3.

As a stand-alone model, the integration of HRM systems lacks the external context emphasised within the earlier models, such as Devanna

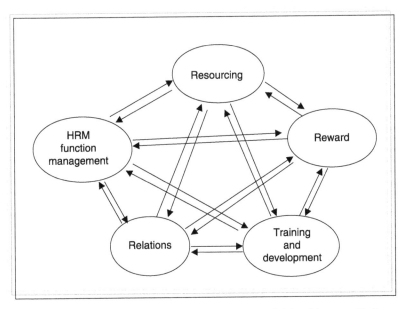

Figure 2.3 Integration of HRM systems (Sparrow and Marchington, © Pearson Education Limited, 1998).

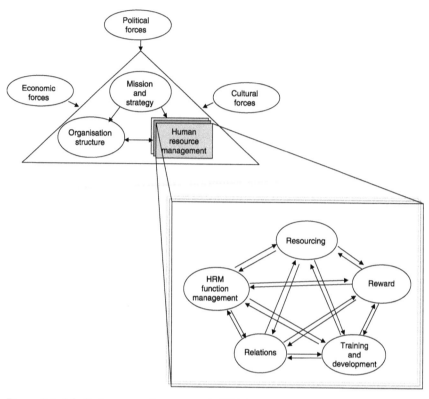

Figure 2.4 A holistic approach to strategic HRM.

et al.'s (1984) matching model. However, used in combination the models provide a clear focus on the demanding operational functions of strategic HRM (see Figure 2.4).

Such an amalgamation of the models incorporates the four types of integration that are necessary for effective strategic HRM: organisational (and environmental) integration, policy integration, functional integration and process integration. According to Mabey and Salaman (1995: 169) these consist of the following:

- organisational (and environmental) integration, where a coherent HR strategy is owned by the Board and accepted by line management, and a willingness to incorporate a HR dimension in important strategic decisions.;
- policy integration, which is concerned with the content of the strategy and the extent to which the resulting policies are coherent;
- functional integration, within which emphasis is placed upon a high-quality HR department in terms of professionalism, number of HR staff to total employee count ratio and representation on the Board;

- process integration concerned with the efficiency and quality of HR processes.

In addition to these prescriptive and analytical models of strategic HRM that characterise the development of the strategic HRM paradigm (the matching model, Harvard model, Warwick model and the integration of HRM systems) two more practice, development and implementation focused approaches have emerged: the 'best-practice' approach and the 'best fit' approach. These are discussed in Section 2.2 together with other recent developments in strategic HRM.

Summary: the conceptual differences between industrial relations/personnel and strategic HRM

The concept of strategic HRM developed from the organisational behaviour and management theories, through the industrial relations and personnel management paradigms, as an integrated approach to people management. The fundamentals of the concept emerged, grew and tightened through these early models. However, throughout the development of the concept, several writers highlighted aspects of strategic HRM to contain notes of contradiction. The works of Guest (1987), Legge (1989), Sisson (1993) and Storey (1992) formed some key contributions to the on-going debate on the differences between industrial/personnel and strategic HRM.

Guest (1987) and Sisson (1993) argue in favour of strategic HRM in that it is concerned with the following:

- an integration of HR policies with business planning;
- a shift in responsibility for HR issues from personnel specialists to line managers;
- a shift from the collectivism of management (trade-union relations) to the individualism of management (employment relations);
- emphasis on commitment, flexibility and quality.

Legge (1989), on the other hand, provides an overtly critical perspective on strategic HRM, finding little difference with the underlying values of personnel management. She argues that organisational constraints may well make a truly integrated approach highly impractical as strategic HRM concentrates on managers and emphasises the key role of line management and the responsibility of top management for managing culture.

In contrast to the approaches of Guest, Sisson and Legge, Storey's (1992) contribution to the debate was in a form of an 'ideal type' classificatory matrix of a 27-item checklist (Table 2.5) for research and analytical purposes. This instrument allows for sets of approaches be pinpointed in organisations by highlighting the main features of each, industrial relations/personnel and strategic HRM, and outlining the differences between the two in an exaggerated way.

Table 2.5 27-point checklist (Storey, 1992)

Dimension	IR/personnel	Strategic HRM
Belief and assumptions		
1 Contract	Careful delineation of written contracts	Aim to go 'beyond contract'
2 Rules	Importance of devising clear rules/mutuality	'Can do' outlook: impatience with rule
3 Guide to management action	Procedures/consistency control	'Business need'/flexibility/commitment
4 Behaviour referent	Norms/custom and practice	Values/mission
5 Managerial task vis-à-vis labour	Monitoring	Nurturing
6 Nature of relations	Pluralist	Unitarist
7 Conflict	Institutionalised	De-emphasised
Strategic aspects		
8 Key relations	Labour-management	Business-customer
9 Initiatives	Piecemeal	Integrated
10 Corporate plan	Marginal to	Central to
11 Speed of decisions	Slow	Fast
Line management		
12 Management role	Transactional	Transformational leadership
13 Key managers	IR/personnel specialists	General/business/line managers
14 Communication	Indirect	Direct
15 Standardisation	High (e.g. 'parity' an issue)	Low (e.g. 'parity' not seen as relevant)
16 Prices management skills	Negotiation	Facilitation
Key levers		
17 Selection	Separate, marginal task	Integrated, key task
18 Pay	Job evaluation: multiple, fixed grades	Performance-related: few if any grades
19 Conditions	Separately negotiated	Harmonisation
20 Labour-management	Collective bargaining contracts	Towards individual contracts
21 Thrust of relations with stewards	Regularised through facilities and training	Marginalised (with the exception of some bargaining for change models)
22 Job categories and grades	Many	Few
23 Communication	Restricted flow/indirect	Increased flow/direct
24 Job design	Division of labour	Teamwork
25 Conflict handling	Reach temporary truces	Manage climate and culture
26 Training and development	Controlled access to courses	Learning companies
27 Foci of attention for interventions	Personnel procedures	Wide-ranging cultural, structural and personnel strategies

In conclusion, despite the diversity of opinion as regards to the differences between industrial relations/personnel and strategic HRM, Storey (1992: 271) points out that strategic HRM carries the potential to bring coherence and direction to a cluster of personnel interventions through an approach which is complete with management techniques and underpinning philosophy. Within the modern business environment, this view suggests that the effectiveness of an organisation largely depends on the efficient use of human resources via practices designed to maximise organisational integration, employee commitment, flexibility and quality of work. This puts forward a comprehensive approach towards the management of people within organisations, which is integrated, individualistic and business focused, and incorporates flexibility. Key development from the earlier models is the concept's concern for factors external to the organisation, the environment. These qualities make the approach a more effective solution for contemporary people management than those grounded in the personnel management paradigm, which focused on procedures and control, administration of employment contracts and job grades and collective bargaining with little strategic involvement.

Although strategic HRM seems a well established field of study with its many models and theories, as identified earlier, it has many critics and sceptical supporters. An ongoing debate on whether strategic HRM is just a new name for the old approach; personnel management, has faithfully stayed on the research agenda ever since the development of the early models. Answers as to the current state of this debate cannot be formulated in a simple manner, and as aspects of the strategic HRM approach to managing the employment relationship contain notes of contradiction, there are no signs for the discussion coming to an end in the near future (see Boxall, 1992; Beardwell and Holden, 1997: 23; Sparrow and Marchington, 1998: 12–13; Purcell, 1999; Marchington and Grugulis, 2000; Gibb, 2001). However, the distinctions drawn here provide a useful analytical lens for examining the practices used by construction firms with respect to their people management policies. In addition, these debates raise several important questions which will be examined later in the book. For example, to what extent have construction companies embraced the individualisation of the employment contract? To what degree has responsibility for the HRM function been devolved to line managers or retained by specialist HR departments? And how do fragmented project-based construction companies ensure a consistency and fairness in the delivery of the HR policies and approaches? These are important questions, the answers to which to some extent define the extent to which construction companies have embraced the strategic HRM paradigm.

2.2 Recent developments in strategic HRM

One of the prominent themes in recent years have been investigation into whether strategic HRM actually delivers high performance. Several studies have investigated the crucial people–performance link that strategic HRM advocates. This has resulted in the adoption of many models and theories which build on strategic management concepts, such as the High Commitment Management (HCM), resource-based view of the firm, High-Performance Work Systems (HPWSs) and best practice and best fit approaches. This section discussed these and their application (where appropriate) to the construction industry. Clearly some theories are more suitable to certain sectors or competitive contexts (Torrington *et al.*, 2005). For example, Guest (2001) argues that HCM/HPWSs might be appropriate for manufacturing where the best fit approach is more realistic in the service sector. Thus, in line with the general philosophy of this book, the aim is not to establish one theory that is right. Rather, focus is on discussing the benefits, criticism and application of these ideas in order to develop suitable foundations for a framework of analysis of the research findings later in the book.

Resource-based view

One of the significant developments in the literature on the relationship between strategic HRM and competitive advantage has drawn on the concept of a 'resource-based view of the firm'. This was initially a strategic management concept which defined the firm as a collection of productive resources with focus on costly-to-copy attributes that are valuable either by taking advantage of environmental opportunities or by offsetting environmental threats (Penrose, 1959: 24; Conner, 1991: 121 and Barney, 1991, in Boxall and Steeneveld, 1999: 444). This produces a form of profitability which rests on superior resource management rather than monopolistic output restriction (Beardwell and Claydon, 2007, provide a comprehensive discussion on this theory). Boxall and Steeneveld (1999) investigated this approach to strategic HRM–competitive advantage relationship through a longitudinal study of engineering consultancies in New Zealand. Their work begun from a standpoint that:

> there is absolutely no need to 'prove' the existence of a relationship between labour management and performance in profit-seeking enterprise. The interesting questions revolve around *how* some firms do it better: engaging and utilizing human talents in ways that deliver more satisfying outcomes for investors, employees and society at large.
>
> (ibid.: 443, emphasis in original)

Although a positive relationship between strategic HRM and competitive advantage was supported, the findings on 'how some firms do it better' were inconclusive since none of the primary subjects in the study had established an enviable form of superiority. The resource-based view of the firm has also been widely criticised as a concept, notably by Priem and Butler (2001) who suggest, for example, that it is flawed as different resources can generate the same value for firms, and hence it is rather prescriptive as a strategy. Nonetheless, it offers a potentially powerful justification for thinking more strategically about the human resource capabilities of the firm. Accordingly, another recent concept, that of 'dynamic capabilities', has emerged which rather than seeing a firm's resources as the key to competitive advantage, sees the capabilities of the firm in being able to reconfigure its routines to respond to changing environments as the key to competitiveness (see Teece *et al.*, 1997; Green *et al.*, 2008). As Green *et al.* point out, this theory is also not without its critics, but this offers an interesting counter position to the resource-based view, as it sees companies' strategic capability as founded on their ability to change and adapt. A shift towards seeing capabilities as dynamic rather than static recognises the importance of learning and development as opposed to knowledge acquisition per se. It therefore has important resonances to the ways in which construction companies manage people.

AMO, high commitment management and high-performance work

The 'People and Performance' model developed by Boxall and Purcell (2003) asserts that performance is a sum of employee Ability + Motivation + Opportunity (AMO). The A, M and O offer a structure for identifying desirable components for a high-performance organisation on the basis that, people perform well when:

- they are able to do so (they *can do* the job because they possess the necessary knowledge and skills);
- they have the motivation to do so (they *will do* the job because they want to and are adequately incentivised);
- their work environment provides the necessary support and avenues for expression (e.g. functioning technology and opportunity to be heard when problems occur) (Boxall and Purcell, 2003: 20, emphasis in original).

Collectively, rigorous recruitment and selection, together with training and development, increase ability levels. Career development, pay, job security, job challenge/job autonomy and work–life balance policies in turn enhance motivation. Finally, communications, employee involvement, teamworking and appraisal ensure employees an opportunity to contribute. Thus, the principles of AMO contend that adopted alone any of these practices are unlikely to affect performance.

One study examined the relevance of this theory in the construction industry. The principles of AMO were applied to data on variables that a sample of managerial and employee respondents in a large construction organisation considered important to take into account in employee resourcing decision-making (see Raidén *et al.*, 2006). The model proved a useful framework of analysis, although a single case is not sufficient to suggest conclusive evidence of applicability in the industry more widely.

The AMO model has served as the foundation for much research on the crucial people-performance link that strategic HRM advocates. However, there seems to be little agreement in terminology and application. High-Performance Work Systems are generally considered the American equivalent to British High Commitment Management, with HR bundles sometimes being used synonymously with both (for further discussion see Huselid, 1995; Wood and de Menezes, 1998; Purcell, 1999; Hutchinson *et al.*, 2000; Boxall and Purcell, 2003). One difference observed by some (e.g. Hutchinson *et al.*, 2000) is their discrepancy in focus: HCM focuses more on the desired outcomes of strategic HRM than the practices themselves like the notion of HPWSs.

In short, HPWSs are built on the AMO model but with an added consideration for employee/organisational outcomes and wider industry/societal context. The concepts were first developed in the US manufacturing context (Appelbaum and Batt, 1994) and later popularised through studies in US steelmaking, clothing manufacture and medical electronics manufacture (Appelbaum *et al.*, 2000). More recently, Boxall (2003) has examined the theory in the service sector and Marchington *et al.* (2003) in small road haulage firms. Purcell *et al.* (2003) focused on the impact AMO had in 12 UK-based organisations that had adopted sophisticated approaches to strategic HRM. In general, their findings suggested that organisations seeking to optimise employee contribution must develop the bundles of strategic HRM policies and practices so that they 'fit' business strategy *and* meet the needs of individuals. This confirms MacDuffies' (1995: 198) hypothesis that the 'HR bundle or system must be integrated with complementary bundles of practices from core business function (and thereby with the firm's overall business strategy) to be effective'. Although they recognise that there needs to be a good fit between HR and operations, Purcell *et al.* (2003) clearly focus on leadership and 'big vision', and their bundled impact on performance in unravelling the 'black box' of people–performance connection. Their research found that the key is finding ways of inducing and encouraging people to exert discretionary behaviour; that is 'to go the extra mile', through these bundles of strategic HRM policies. However, policies need effective implementation and in that, line managers' role is central. Leadership influences motivation, commitment and satisfaction through fair communication, support and respect. The big vision relates to the organisational vision, values and culture; a direction from the top that leads to unified action throughout the organisation.

One of the Purcell *et al.*'s (2003: 33) case organisations was particularly relevant to the construction sector, where project teams formed the unit of analysis and the organisational structure was built upon a series of communities. These included short-term operational communities, or project teams, which work for a particular client; vocational communities that support specialists across the project teams; and non-hierarchical staff communities which form the principal means of internal communication. Together with the organisational culture, which was based on making the organisation 'a good place to work', the matrix project structure provided the organisation with 'the ability to combine all aspects of human resource and organisational processes together and build a strong value-based organisation' (ibid.). This resulted in the professional employees 'knowing what was going on, having a say in organisational decisions that affected their job or work, working in a supportive team environment and doing challenging work that helped develop their career skills' and thus enjoying high job satisfaction and contributing above average performance for the benefit of sustainable competitive advantage for the organisation (ibid.: 34–35).

Despite useful application of HPWSs/HCM in other project-based sectors, Druker (2007) notes that there is little evidence of the adoption of such practices in the construction industry. Indeed, she goes on to suggest that in the current climate, involving whole workforce in sustained improvement, learning and no-blame culture based on mutual interdependence and trust may be impossible to achieve.

Best practice vs best fit

In addition to the models derived from strategic management, and the prescriptive and analytical models of strategic HRM discussed in Section 2.1 (the matching model, Harvard model, Warwick model and the integration of HRM systems), two more practices, development and implementation focused approaches have emerged: the 'best-practice' approach and the 'best fit' approach.

The best-practice approach is based on the belief that adopting a set of best HRM practices leads to a superior organisational performance. Pfeffer (1994) lists a set of seven 'best HRM practices':

- employment security
- selective hiring
- self-managed teams
- high compensation contingent on performance
- training
- reduction of status differentials
- sharing information.

These reflect the principles of HCM, but this approach has received heavy criticism on its notion 'one size fits all'. Indeed, organisational HPWSs are highly idiosyncratic and must be tailored carefully to each firm's individual situation to achieve optimum results (Armstrong and Baron, 2002). It is for this reason that best fit is seen more effective than best practice. 'Good practice' and 'leading-edge practice' are accepted as useful indicators of the type of solutions that work in certain situations, but a universal prescription of a one right way must be impossible. The best fit approach encourages continuous analysis and evaluation of the business and HR needs within the organisational contexts (culture, structure, technology, processes, environment) and suggests that in response to the outcomes of such an evaluation a selection of good/leading-edge practices should be implemented. Nevertheless, the best fit approach has also been heavily criticised in that it is limited by the impossibility of modelling all the contingent variables, their interconnection and the way in which they influence each other (Purcell, 1999). Indeed, Purcell (1999) argues for being more sensitive to the process of organisational change and avoid being trapped in the logic of rational choice.

At present, this is one of the most significant academic controversies in the field of HRM, as well a managerial issue with significant implications to many organisations (Torrington *et al.*, 2005).

2.3 Summary

This chapter has provided a general review of the historical development of theories and practices of managing people in organisations, with some specific references to the construction industry. An outline of the major theories in organisational behaviour was followed by discussion on industrial relations, personnel management, HRM and finally, the strategic HRM. This review provides a platform of knowledge on which a more specific analysis is provided in Chapter 3.

3 Components of strategic HRM

Strategic HRM comprises three distinct but interrelated functions: employment relations, learning and development, and employee resourcing. Each has its definite focus but the boundaries between the functions can overlap significantly. In short, employment relations provide an overarching management philosophy for the management of people within an organisation. Learning and development centre on training and development of people and organisations. Employee resourcing seeks to bring appropriate people into the organisation, manage their performance, look after the related HR administration, handle their exit from the organisation and reflect change through the various processes the function involves. This chapter discusses these three key components in the context of the construction industry with particular emphasis on employee resourcing.

3.1 Employment relations

Employment relations provide an overarching management philosophy, or style, for the management of human resources within an organisation. The style managers adopt for dealing with their staff is of significant importance to resourcing decision-making since a key feature of strategic HRM is to devolve much of the HR responsibility to the operational line management (Sparrow and Marchington, 1998; Pilbeam and Corbridge, 2002). This requires careful management of the HR-line interface if organisations are to maintain a healthy balance of interests between the employer and its employees. Thus, the HR professionals' role in advising the operational managers, for example on the increasingly complex web of national and EU legislation remains crucial. The extent of this task is dependent on the management style held within an organisation, since it influences the organisational culture, determines the approach adopted towards conflict resolution and dictates the kind of strategies and practices that are likely to succeed within an organisation.

Gennard and Judge (2002: 208–10) summarise different management styles falling under the broad categories of unitary or pluralist, and within these authoritarian, paternalistic, consultative, constitutional or opportunist approaches. Each combination has two additional dimensions: individualism and collectivism. Purcell and Ahlstrand (1994) developed a useful model for classifying management styles along the individualist–collectivist dimensions. This is shown in Figure 3.1.

This model defines the two dimensions, individualism and collectivism, as interconnected. Hence, they are incorporated within a matrix structure. The individualistic dimension (on the vertical axis) places the management styles on a continuum where on the 'resource' end employees are viewed as individuals with needs, aspirations, competencies and particular skills of their own and the 'cost minimisation' approach treats them as homogeneous groups of people with personnel policies unable to distinguish between individuals and individual performance. Between the two extremes falls paternalism, which refers to use of personnel policies that emphasise employee loyalty achieved via generous pay and benefits packages.

On the horizontal axis, the collectivistic dimension focuses around employee groups and teams. On the one hand, 'the co-operative' recognises that employees have the right to form themselves into independent or quasi-independent organisations and actively encourages employee participation. On the other hand, 'the unitary' approach avoids any form of collective organisation. This is viewed as a threat to managerial authority. The central category, 'adversarial' approach, focuses on bargaining as the key activity, in which each party seeks to restrict information flows in order to reach a compromise agreement.

A combination of different levels of the two dimensions result in six distinct approaches to employment relations: traditional, paternalistic, sophisticated human relations, bargained constitutional, modern paternalistic and

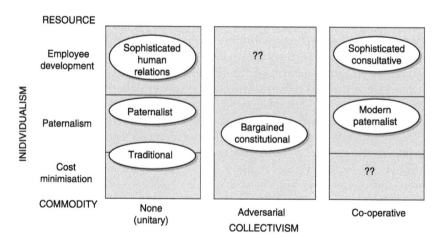

Figure 3.1 The management style matrix (Purcell and Ahlstrand, 1994: 178).

sophisticated consultative. Purcell and Alhstrand (1994: 188–201) describe these as follows:

- *Traditional* – aim to minimise costs of employee remuneration, recruitment and training, no mechanism for recognising employee views, 'command-and-control' management, common result high staff turnover.
- *Paternalistic* – stability and order, caring image, reasonable pay, little expectations or opportunities for promotion, training provision centred on matching the skill requirements of the current task/job.
- *Sophisticated human relations* – commitment to employee involvement and teamwork, aim to generate employee loyalty and commitment and thereby maximise productivity and responsiveness to change, carefully formulated and conducted HRM procedures, above average pay rates, recognition for each individual employees contribution to the organisation, expectation for employees to work 'beyond their contract'.
- *Bargained constitutional* – employees managed much in the style of 'traditionalists' or 'paternalists' but with the difference that employee organisations (such as trades unions or works councils) are recognised, management is unlikely to value union presence, policies developed around the need to achieve stability and control.
- *Modern paternalistic* – welfare-orientation, employee organisations recognised and valued, focus on establishing constructive relationships, management of change.
- *Sophisticated consultative* – almost identical to the 'sophisticated human relations' approach in its heavy investment on employees in order to maximise their contribution to the organisation, but with the difference that employee representation through collective means is actively encouraged, aim to create constructive relationships.

The '??' boxes in the figure represent inherently unstable conditions which are unlikely to last. For example, it is unrealistic to assume that 'cost minimisation' characterised by lowest possible wage levels and organisational reluctance to engage in training and developmental activities would be met by highly co-operative behaviour on the part of an employee welfare organisation (such as trades union or works council).

Traditionally in construction, managers from a craft or engineering background naturally attempted to achieve 'the best' technical/financial business outcome and took appropriate steps to implement this (De Feis, 1987). Many construction managers still operate in this fashion, focusing on the achievement of financial, programme and quality outcomes over other project performance criteria (see Dainty *et al.*, 2003a). Indeed, Druker (2007) relates the industry's approach to risk management to its approach to industrial relations. Essentially, organisations in the industry preserve collective bargaining as a risk management strategy without affording workers the benefits of voice. Collective bargaining is normally designed to do just this: provide employees with voice (Druker, 2007, provides a comprehensive overview of the institutional framework for collective bargaining in the UK construction sector). Thus, they have no route to rehearse grievances or contribute to change at the level of industry. This is particularly problematic because of the large numbers of self-employed personnel working on site because their status does not allow they voice within organisational (employment) grievance procedures either. Organisationally, this has led to little by way of innovation in managing industrial relations within the industry, coupled with no take-up of HPWSs/HCM as discussed earlier (Druker, 2007). Thus, not surprisingly, much of the current project allocation decision-making tends to be ad hoc and based on the implicit knowledge of senior managers (Raidén *et al.*, 2002a, 2002b). The skills and knowledge requirements of the project dominate the decision-making, at the expense of individual needs and aspirations. Employee resourcing decision-making is at an imbalance, which leads to employee dissatisfaction and staff turnover. In conclusion, Druker (2007) suggests that the industry operates within the 'traditional' view: aim to minimise costs of employee remuneration, recruitment and training, no mechanism for recognising employee views, 'command-and-control' management and common result high staff turnover.

Staff turnover may also be the consequence of an organisation not supporting employee work–life balance (this is discussed in detail under employee resourcing, Section 3.3). Other factors connected with staff turnover include limited career/development prospects and lack of management attention to employees through feedback and provision of opportunity to influence decisions (Afifi, 1991; Dainty *et al.*, 2000b). The consequences of high staff turnover can be serious in terms of increased costs of recruitment and training, interruptions to work flows and lower staff morale which in turn can affect team and organisational performance (ACAS, 2003). Some staff turnover, however, helps bring in new ideas into the organisation in balance with the continuity of existing operations. The Chartered Institute of Personnel and Development (CIPD, 2001) estimate a healthy rate for staff turnover to be around 10%. Figures for construction organisations are often recorded to double that (e.g. 21.9% in 2001 and 21.7% in 2006; CIPD, 2002 and CIPD, 2007, respectively). These figures look unhealthy, particularly when almost three quarters (71%) of the CIPD's annual recruitment, retention and turnover survey respondents reported that the level of staff turnover had a negative effect on their performance (CIPD, 2007: 31). Furthermore, a significant 15% of this group stated that the effect was serious.

Overall, organisations facing high levels of staff turnover often report it negatively affecting their performance. In contrast, organisations with a staff turnover of 10% or less are likely to report it as having a positive or no effect on their performance (CIPD, 2003b). Thus, bearing in mind the dynamic, geographically dispersed project-based characteristics of the industry, the effects of a 21.9% (high) staff turnover are likely to cause considerable disruption to the team and organisational performance within the large construction contractors (Chapman, 1999; Love *et al.*, 2002).

The communication imperative

Effective communication is a crucial enabler of performance for any business. However, within a transitory and fragmented industry like construction, with its itinerant and largely self-employed workforce, effective communication is extremely problematic. Everyone involved in construction plays a part in a complex communication network. Seeing the project environment as an interconnected network of actors is appropriate because every such venture, no matter how small or well defined, can be successfully completed without interactions between people within (and between) organisations (Dainty *et al.*, 2006). Because it is project-based, its groups and networks are temporary in nature. Hence, relationships and interactions continually change to reflect the dynamic nature of the workplace. Hence, there is always an element of uncertainty with the potential to undermine communication to the detriment of project performance.

The importance of communication to organisations is succinctly summarised from an HRM perspective by Armstrong (2001: 807):

- *Achieving coordinated results* – organisations function by means of the collective actions of people, but independent actions lead to outcomes incongruent with organisational objectives. Coordinated outcomes therefore demand effective communications.
- *Managing change* – most organisations are subject to continuous change. This, in turn, affects the employees that work within them. Acceptance of and willingness to embrace change is likely only if the reasons for this change are well communicated.
- *Motivating employees* – the degree to which an individual is motivated to work effectively for their organisation is dependent upon the responsibility they have and the scope for achievement afforded by their role. Feelings in this regard will depend upon the quality of communications from senior managers within their organisation.
- *Understanding the needs of the workforce* – for organisations to be able to respond effectively to the needs of their employees, it is vital that they develop an efficient channel of communication. This two-way channel must allow for feedback from the workforce on organisational policy in a way that encourages an open and honest dialogue between employees at all levels and the top-level managers of the organisation.

Hence, the corollary of poor communications for an organisation is that employees will misread management decisions or react to them in a way that was not intended. Similarly, managers will misunderstand the needs of employees and will therefore suffer from lower performance and a higher turnover of staff. Another requirement for effective communication in construction stems from the industry's propensity to undergo change and transition. Coping with change is more problematic in traditional industries like construction, which have shown a reluctance to embrace new ways of working, but is arguably more important given the disparate pools of knowledge that must be combined within construction projects. In the past, a 'silo' like mentality has prevailed which has been shown to impede knowledge sharing within the industry (Dainty *et al.*, 2004a). However, effective communication has the power to break down such barriers by bringing people together, thereby propagating improved collaboration and integrated working within the sector. Thus, effective communication can be seen as a key enabler of future industry improvement.

The increasing acknowledgement of the importance of communication in organisations is reflected in the rising prominence of communications within the HRM literature. According to Torrington and Hall (1998: 114), upwards communication is vital for the following:

- understanding employees' concerns;
- keeping in touch with employees' attitudes and values;
- alerting managers to potential problems;
- providing managers with workable solutions to problems;
- providing managers with information necessary for effective decision making;
- encouraging employees to contribute and participate with organisational decision making, thereby improving motivation and commitment to organisational actions and directions;
- providing feedback on the effectiveness of downwards communications.

Although construction has long been criticised for its outmoded approach towards many aspects of the HRM function (Loosemore *et al.*, 2003), elements of this approach can be seen to have permeated HRM in the industry. For example, whereas in the past information flow has been mediated by employee representation or trade unions, contemporary approaches have placed an emphasis on communicating directly with the workforce via individualised employment contracts. Indeed, the decline in the significance of collective bargaining is likely to lead to an increased emphasis on direct communication in the future (Emmott and Hutchinson, 1998). Thus, strategies for communicating within construction organisations tend to reflect the new employment paradigm and the shifts in workplace culture that this has engendered. This new communication orthodoxy will be revisited in later sections of this book.

3.2 Learning and development

Learning and development are vehicles for facilitating organisational and individual improvement through training and development. Systematic as well as ad hoc development programmes help to ensure staff have the skills required for their current roles and can develop those required for future posts. It can also work as a motivating factor: significant training indicates commitment to people and the recipients are more likely to feel valued (Sisson and Storey, 2000). From a business point of view, learning and development can be seen as a tool for creating sustainable competitive advantage (Burden and Proctor, 2000).

There are 'soft' and 'hard' elements to learning and development (El-Sawad, 2002: 286). The soft components imply investment in people; the hard aspects suggest cost and expendability. In practice, the hard elements of learning and development often relate to mandatory training courses, such as health and safety updates. The soft dimensions take a more holistic view within the concepts of organisational learning, continuous development and learning organisation. These frequently operate simultaneously and are mutually supportive, and this indeed may be the most beneficial way of managing learning and development.

Training

Training includes a range of formal and informal activities that are aimed at providing employees with the skills required to carry out their job. This includes the maintenance and further development of their existing capabilities as well as the learning of new competencies (Bratton and Gold, 2003). Training activities (the 'hard' element of learning and development) usually refer to employer-driven, short-term courses focused on the present needs (Myers and Kirk, 2005). In contrast, developmental activities are often more unstructured and incorporate learning undertaken outside the organisational boundaries. Organisational development is the result of collective learning within the members of the organisation.

The training process involves a planned, systematic and cyclical approach for identifying and suitably responding to individual and organisational training and development needs. The first stage in the systematic process consists of *a training needs analysis* (TNA). This can be carried out at an organisational, job/occupational and/or individual level. A comprehensive TNA requires a broad range of internal and external data available for analysis, for example, internally: the organisational strategy, structure, technology, managerial style, culture, working conditions, employees' aspirations and skills; and externally: political, economic, social, technological, legal and competitive environmental information. The assessment of this data ideally leads to measures of (i) current and future business performance needs and (ii) current performance and capability levels; and by comparing the two (iii) to the identification of the current and potential future capability gaps and (iv) which training and development interventions might effectively address them within the (v) target population (McClelland, 1993).

On the basis of the TNA, the actual *training design and delivery* should focus on what is to be learned and how people learn. These influence the choice of appropriate training methods, for example, whether traditional packaged classroom learning experiences are to be offered, computer-based e-learning encouraged or informal, 'organic', on-the-job learning facilitated via mentoring/coaching. Kolb's (1996) learning cycle and Honey and Mumford's (1982) learning styles inventory provide useful frameworks for the design and delivery of training and development solutions that support the achievement of the desired learning outcomes. The traditional 'reinforcing theory with examples' does not work with every individual, group, situation or even topic, therefore different 'learning modes' and 'learning styles' should be accommodated. Kolb's (1996) learning modes support development via a four-stage process which begins with concrete experience, leads to reflective observation and abstract conceptualisation before closing with active experimentation. He argues that such a process promotes both active and passive, concrete and abstract learning and thus delivers quicker and easier learning outcomes. Honey and Mumford (1982) focus on four learning styles. Their instrument comprises activists who learn best by active involvement, reflectors who learn best by reviewing and reflecting, theorists who learn best when new information can be located within the context and concepts and theories, and pragmatists who learn best when they see a link between new information and real-life problems and issues.

Finally, a major objective of *training and learning evaluation* is to demonstrate the impact of investment the activities undertaken. Reid *et al.* (1992) suggest five levels at which the evaluation may be useful:

- reactions of trainees to the training programme;
- whether trainees learned what was intended;
- learning transfer back to the work environment;
- whether the training has enhanced departmental performance;
- the extent to which the training has benefited the organisation (the ultimate level)

Fellows *et al.* (2002: 131–6) identify three specific vehicles of training suitable for managerial and professional staff (focus of the research reported here) within the construction industry: professional development, management development and the use of learning networks. The professional development refers to graduate training towards achieving a corporate membership of a chartered institution, such as the Chartered Institute of Building (CIOB), Royal Institution of Chartered Surveyors (RICS) or Institution of Civil Engineers (ICE). Management development seeks to develop the employees' managerial skills beyond the professional membership requirements. This commonly includes formal education and training activities as well as informal, incidental and opportunistic learning Langford *et al.*, 1995; Druker and White, 1996; Fellows *et al.*, 2002). The ICE (2001), for example, have developed guidelines for the organisational management development activities, which also provide useful information for the construction professionals as to their role in the process. However, the number of organisations undertaking management development within the industry is small, although those organisations that do undertake management development tend to place a lot of emphasis on it and support formal technical training courses with coaching. Loosemore *et al.* (2003: 257–258) suggest that the low uptake stems from the following:

- the assumption that training delivery is expensive;
- clashes with production objectives;
- legislative training requirements dressing additional activities as unnecessary add-ons or luxuries;
- staff turnover concerns in relation to the belief that training and developing employees will make them more attractive to other companies;
- a macho environment, within which the traditional classroom education is often seen as a non-productive feminine activity and associated with failure;
- a 'learn on the job' culture.

The third technique, use of learning networks, includes, on the one hand, formal organised networks, which are often based around professional groupings and managed by appropriate professional bodies, and on the other, more informal networks, which develop as companies work together. Supply chains and partnering commonly encourage this type of learning and thus facilitate inter-organisational transfer of knowledge (Barlow and Jashapara, 1998).

Jashapara (2003) researched the impact of learning to organisational performance within construction organisations. The conclusions of the study suggested that the dynamics of competitive forces evident within the industry imply a need for construction organisations to focus their organisational learning on efficiency and proficiency to achieve competitive advantage. To this effect, the industry practice could be argued as suitable and appropriate since much of the training focuses on meeting statutory requirements and providing staff with the skills required to carry out the tasks involved in their current roles (Druker and White, 1996; Chan *et al.*, 2001; Eckford *et al.*, 2001). However, a short-term focus of organisational learning on efficiency and proficiency undermines the long-term individual career development and organisational succession planning benefits that potentially follow from strategic learning and development policy, which takes into account the needs of the organisation *and* the people it employs (Dainty *et al.*, 2000b). In line with the industry's low take-up and commitment to Investors in People

(IiP) initiative, learning organisations and learning and development in general, Hancock *et al.* (1996) found no significant incidence of learning and development in large construction companies. Dainty *et al.* (2000b) recognised the missed opportunities of strategic learning and development and suggested a fundamental realignment of the function with the employee needs so that maximum benefits, effective recruitment and retention and competitive advantage could be achieved.

This operational/mechanistic view of learning and development falls under the 'hard' category. The more developmental 'soft' side clearly highlights the attitudinal aspects of learning and development, as demonstrated within Organisational Learning.

Organisational learning

Organisational learning forms the 'ideal' type of learning and development within a classification of four models-in-practice (El-Sawad, 1998a: 227) illustrated in Table 3.1.

Organisational learning fosters change and renewal on a continuous basis, and encourages creativity and innovation. It seeks to continuously question the norms, which define effective performance. Thus, continuous development is a central element of organisational learning. It emphasises the attitudinal dimension of learning, which is clearly reflected in the CIPD statement on continuous professional development: '[it] is an attitude as well as a process – the continual and conscious search for, and recognition of, learning in almost every activity and situation' (El-Sawad, 2002: 295).

Chan *et al.* (2005) highlights the many gaps in our knowledge about organisational learning in construction. In general, the organisational learning approach to learning and development recognises values and positively encourages people to take advantage of learning opportunities on-the-job, off-the-job and outside of work. Ultimately, this leads to a climate of self-development, which in turn supports the concept of the learning organisation (see hereafter).

Learning organisation

Pedler *et al.* (1988) define learning organisation as 'an organisation which facilitates the learning of all its members and continuously transforms itself ... [It is a] vision for an organisational strategy to promote self-development amongst the membership and to harness this development corporately by continuously transforming itself as part of the same process'.

A learning organisation is characterised by learning climate, ethos of self-responsibility and self-development, learning approach to strategy, participative policy-making, internal collaboration, continuous development, reward flexibility, inter-company learning, and temporary structures responsive to environmental changes, which suggest an extremely attractive

Table 3.1 Learning and development models-in-practice (after El-Sawad, 1998a: 225–226)

	Intermittent pattern	Institutionalised pattern	Investor pattern	Internalised pattern
Managerial commitment	Low	Apparent	High	Very high
Training and development activity	Little visible activity	High level of visible activity	Systematic, cyclical, organisationally-managed approaches to identifying and responding to development needs	Acceptance of a strong developmental ethos (but quietly so), activities more visible
Training and development interventions	Infrequent, ad hoc, reactive, often in response to a crisis	Large budgets invested in extensive off-the-job, fixed menu training on the basis of assumed needs	Substantial expenditure carefully managed, prioritised and targeted at actual business-defined development needs	Developmental philosophy strongly embedded within the organisational culture, learning a day-to-day business-as-usual activity
Learning	No organised learning	Disorganised learning	Organised learning	Organisational learning

working climate (Newell, 2001). Indeed, most successful organisations that incorporate learning and improvement as an integral aspect of their organisational culture are as follows:

- *Strategically led*, with all employees able to articulate the vision with understanding.
- *Competitively focused*, driven by the need to compete at the highest levels and well aware of what competition is up to.
- *Market-oriented*, close to both the customer and the consumer, via research into their changing needs and the agility to respond quickly.
- *Employee-driven*, with highly competent people united by the desire to learn, innovate and experiment.
- *Operationally excellent*, with finely tuned processes and clear performance measures (Newell, 2001: 111).

Since Senge's (1990) and Pedler *et al.*'s (1991) initial work, the concept has taken a central role in discussions which focus on the more advanced approaches to learning and development (Mumford, 1995; Garavan, 1997; Stewart, 2001; Johnson, 2002; Phillips, 2003; Nyhan *et al.*, 2004). The detail has expanded to cover a range of more specific areas, such as single- and double-loop learning, transformational and adaptive learning, the learning process and systems thinking.

Single/double-loop learning and adaptive/transformational learning

Single- and double-loop learning refer to different hierarchical levels of learning within an organisation. In single-loop learning, errors are detected and corrected in a 'continuous improvement' process through incremental or adaptive learning (Stewart, 2001). Double-loop learning demonstrates a deeper level questioning and challenging of the organisational success formulas. This represents transformational learning, which seeks to introduce radical change. Much of the focus in research builds on the work of Senge (1990), who emphasised the advantages of double-loop learning over single-loop learning. However, some (e.g. Nevis *et al.*, 1995; Nicolini and Meznar, 1995; Appelbaum and Goransson, 1997) argue that this takes an unnecessarily narrow view and suggest that an approach which combines both single- and double-loop learning perspectives provide a more balanced outlook towards learning and development (Appelbaum and Goransson, 1997).

The learning process

Theories relating to the process of learning emphasise the continuous nature of learning. Kolb's (1984) learning cycle is perhaps the most established descriptive model of individual, team and organisational learning. This explores the cyclical pattern of four stages in learning: experience, reflection,

Table 3.2 Organisational learning as construct of sub-processes (developed from Huber, 1991)

Sub-process	Sub-sub-processes	
1 Knowledge acquisition	1.1 Drawing on knowledge available at organisation's birth	
	1.2 Learning from experience (Kolb's learning cycle)	1.2.1 Experience 1.2.2 Reflection 1.2.3 Conceptualising 1.2.4 Action
	1.3 Learning by observing others	
	1.4 Drawing on external sources	
	1.5 Collecting and using information about organisational performance	
2 Information distribution		
3 Information interpretation	3.1 Framing 3.2 Cognitive maps	
4 Organisational memory	4.1 Organisational culture	

conceptualising and finally action. Huber's (1991) construct of organisational learning constitutes of four sub-processes, which in turn include further sub-sub-processes (see Table 3.2).

Nevis *et al.* (1995) suggest a similar knowledge-based structure for organisational learning process. This consists of three stages: knowledge acquisition, knowledge sharing and knowledge utilisation. However, Nicolini and Meznar (1995) and Appelbaum and Goransson (1997) argue that both models refer to the cognitive processes of learning that take place in organisations and thus constitute only one aspect of organisational learning. The other aspect is social construction of organisational learning. This refers to the self-reflective process involved in transforming cognitive learning into abstract knowledge. It also refers to the symbolic and political processes through which organisational leaders develop their identity.

Systems thinking

Systems thinking provide a methodology for understanding organisations as a whole by exploring their patterns and the nature of interrelationships. Senge's (1991, 1994) seminal works about systems thinking and its application to organisations introduced the idea as the basis for developing a learning organisation. He identified five 'component technologies' that are crucial for such organisations:

- *Personal mastery* – continually clarifying and deepening our personal vision, of focusing our energies, of developing patience, and of seeing reality objectively.
- *Mental models* – deeply ingrained assumptions, generalisations, or even pictures or images that influence how we understand the world and how we take action.
- *Building shared vision* – the capacity to hold a shared picture of the future we seek to create.
- *Team learning* – teams, not individuals, are the fundamental learning unit in modern organisations.
- *Systems thinking* – focus on how the thing (an organisation or an aspect of an organisation) being studied interacts with the other constituents of the system of which it is a part.

These elements (single/double-loop and adaptive/transformational learning, the learning process and systems thinking) have been used to construct frameworks of analysis for determining the extent to which organisations adopt/ implement learning organisations in practice. Two interesting recent developments include a model of understanding the dimensions of organisational learning by Nyhan *et al.* (2004) and the concept of a 'chaordic enterprise' (van Eijnatten and Putnik, 2004).

Model of understanding the dimensions of organisational learning

Nyhan *et al.* (2004) suggests that a learning organisation exhibits four characteristic features:

- coherence between the formal organisational structure and informal culture; and organisational goals and individual employee needs;
- challenging work;
- support and provision of opportunities for learning;
- partnership between vocational education, formal training and informal learning and development.

Their hypothesis suggests that the key to becoming a learning organisation *lies in the capacity to understand and see how the different and often seen as opposing dimensions of organisational life can be reconciled* (ibid.: 75). These contrasting demands are represented along two continuums (Figure 3.2). The horizontal axis represents, at one end, the need to formalise and make transparent and on the other to manage the informal organisational culture. The vertical axis represents, at one extreme, the need for learning and development strategies that support the organisational performance objectives and on the other encouragement of personal responsibility in meeting employee needs.

The contrasting demands in this model pose constantly changing forms of

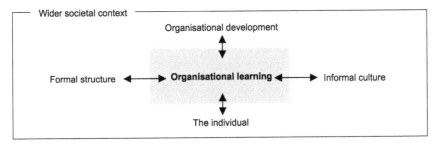

Figure 3.2 Model of understanding the dimensions of organisational learning (Nyhan *et al.*, 2004, with permission from Elsevier).

conflict for organisations and so a linear 'either–or' approach is rejected (Nyhan *et al.*, 2004: 77). The diverse range of challenges faced by organisations in the modern business environment requires managers to respond to each situation on its individual merits. Structural procedures may work well to solve one type of conflict, whereas another situation may benefit from more informal cultural guidance. The rejection of the linear 'either–or' approach also implies a need to pay attention to both the organisational development requirements and the individual employees' needs for training and advancement. Accordingly, more inclusive 'both–and' approach is put forward as appropriate for a learning organisation.

The concept of 'chaordic enterprise'

A second recent development in the field is the 'Chaordic enterprise'. This originated from chaos and complexity theories, which accept the feature that people act upon a system of which they themselves are an inseparable part' (van Eijnatten and Putnik, 2004: 423). 'Chaord' derives from *cha*-os and *ord*-er. 'Chaordic' refers to anything simultaneously orderly and chaotic . . . existing in the phase between order and chaos. The combination of 'chaordic' and systems thinking produced 'chaordic system': 'an entity in which nothing ever happens quite the same twice, but enough happens in a tidy enough way to preclude complete pandemonium'; and further, the 'chaordic enterprise' (Fitzgerald and van Eijnatten, 1998: 264). A Chaordic enterprise can therefore be defined as:

> . . . [an] enterprise in which the two most fundamental properties of reality [chaos and order] are maintained in dynamical balance by virtue of an intentional process of management.
>
> (ibid.)

Key features of a chaordic enterprise include discontinuous growth, organisational consciousness, connectivity, flexibility, continuous transformation and self-organisation. van Eijnatten (2004) explains these as follows: The discontinuous growth refers to the cyclical nature of organisational development from birth to growth, stability, decline and instability through to growth again. Development and learning are seen as discontinuous in this process. The organisational consciousness places importance on organisational mind (collective vision) as the driving force for change. Connectivity emphasises the nature of an organisation as a whole, and a part of a wider system. Flexibility in a chaordic enterprise signifies the fact that future is unpredictable. Consequently, organisational focus should be on preparing for change, not planning for change, and the how is to be made up as situations arise. Continuous transformation refers back to the cyclical nature of organisational development from birth to growth, stability, decline and instability through to growth again. According to this element of a chaordic enterprise, organisations should build mechanisms that enable them to initiate change very early on in decline in order to avoid steep falls. However, it is recognised that rebuilding organisations from states of deep instability often creates novel new forms. Finally, self-organisation refers to the need for a collective vision that is shared by all and thus directs all thought and action.

Some believe the chaordic theories will emerge as the principal science of the next century in studying the complex, non-linear, adaptive systems, which modern organisations present, particularly in industries like construction which are subject to a wide range of competitive and regulatory forces which lie outside of the control of the individual firm. Accordingly, the chaordic enterprise is suggested to provide an appropriate conceptual framework for using complexity to understand organisational patterns and human interactions in learning organisations.

The prevalence of learning organisations in construction

Despite its attractive qualities, the concept of learning organisation has received minimal attention at an applied level within the construction industry (Loosemore *et al.*, 2003: 255). Druker *et al.* (1996) found construction organisations being far from learning organisations. The industry is known for its low take-up of the IiP initiative and poor commitment to learning and development. Kululanga *et al.* (1999), Ford *et al.* (2000) and Loosemore *et al.* (2003) provide recent research evidence to support this, and suggest one possible reason for the low take-up and commitment to learning and development being the predominance of an engineering culture that focuses on technology instead of people. Loosemore *et al.* (2003: 257–258) suggest that the low commitment to learning and development is rooted in the assumption that training delivery is expensive and clashes with production objectives and legislative training requirements. This style of thinking dresses additional activities as unnecessary add-ons or luxuries, and views staff turnover concerns with a belief that training and developing employees will make them more attractive to other companies. Finally, within the macho environment traditional classroom education is often seen as a non-productive and there is a prevalence of the 'learn on the job' culture. These factors combine to reduce the significance of learning organisation in many construction firms, despite its importance to the development of dynamic capabilities (see aforementioned). Love *et al.* (2000) provide a discussion on the possibilities for learning organisations in construction through a conceptual framework.

3.3 Employee resourcing

The major components of employee resourcing are staffing, performance, HR administration and change management (Taylor, 2005). These main functions consist of several individual, but interrelated management activities, which are summarised in Table 3.3 and discussed hereafter under headings extracted from the table. Important current themes and key issues are highlighted in the discussion. Much attention has been paid to the construct of psychological contact, flexibility, careers, employee involvement, adoption of technology and work–life balance in recent years (Sparrow and Cooper, 2003; Torrington *et al.*, 2005; Taylor, 2008). Such issues have significant impact on how the resourcing function is managed.

The staffing and performance objectives aim to ensure that the right number of employees with the right skills and competencies are in the right place at the right time. This can be conceptualised as a 'balancing act' in which managers must take into account the longer term strategic considerations of human resource planning while providing immediate solutions for the shorter term operational issues, such as recruitment and selection, team deployment, dismissal and redundancy. Ideally, management of staffing and performance will be simultaneously concerned with ensuring that performance is achieved while facilitating employees' career progression and offering them appropriate reward for their efforts. In order to support the information flows required, the HR administration function focuses on the collection, storage and use of employee data. The change aspect of the function aims to facilitate the continuous evolution of the organisational strategies and practices through the interrelated aspects of staffing, performance and HR administration.

Human resource planning

The primary concern of human resource planning (HRP) is to integrate the strategic and operational requirements of the business with a workforce equipped to provide the services and products that customers demand. Some experts doubt as to whether HRP is a worthwhile activity given the turbulence of modern business environment (Marchington and Wilkinson, 2002). Notwithstanding this, planning is crucially important, especially within the dynamic project-based sectors, in that it can help reduce uncertainty,

Table 3.3 Employee resourcing tasks with related HRM activities and objectives (after Taylor, 2005)

Strategic HRM objective	Strategic HRM activity	Tasks involved
Staffing	Human resource planning (HRP)	Strategic human resource forecast – an input; development of a human resource plan – an output
	Recruitment and selection	Identification and analysis of recruitment needs, drawing of job descriptions and person specifications, advertisement of the vacancy, shortlisting candidates, selection process utilising appropriate selection techniques (i.e. interviewing, assessment centres, etc.), selection of the 'right' candidate, induction
	Team deployment	Formation and building of effective teams, deconstruction and redeployment of teams
	Exit	Redundancy, retirement, dismissal, voluntary exit
Performance	Performance management	Continuous evaluation and performance appraisal; feedback and reward
	Career management	Promotion; personal and professional development planning
HR administration	Collection, storage and use of employee data	Utilisation of appropriate HR administration system, e.g. manual filing system or a computerised human resource information system (HRIS)
Change management	'Change agent'	Ensuring proper recognition is given to significance of change; management of business and strategic HRM processes via which organisational culture and structure continually evolve

introduce structure and create order and action. A human resource plan is developed to act as a means to achieve strategic HRM targets, and thus forms the output (Turner, 2002).

The early literature on construction HRP by Huang *et al.* (1996), Anderson and Woodhead (1987) and Schaffer (1988) dealt with a quantitative approach termed 'manpower planning'. This focused on forecasting the numerical supply and demand of manpower and reconciling these. Methods to carry out the supply analysis included fractional flow models and renewal models. The demand side techniques included time series models, workload methods, regression models and the manpower system model. Schaffer (1988) begun to introduce softer aspects to HRP literature by encouraging organisations to make, what he called *the moral contract*. This was based on interaction, agreement and mood of commitment within the parties involved and simple procedures of operation.

Maloney (1997) built on this type of approach and highlighted the importance of strategic planning for HRM in construction. He noted the current HRP/HRM strategies being largely emergent. In response, he strongly advocated the need for HRM strategies to become more deliberate. Areas of particular importance were identified to include strategic vision, view of human resources, management vs worker orientation, short-term versus long-term orientation and availability of a skilled workforce. Smithers and Walker (2000) added to this the need for increasing planning effectiveness and decreasing the chaotic nature of a project. Their commentary on the motivation problem within the industry identified a need

> for construction companies to more strategically target their workload toward a more profitable work while improving the construction site management style.
>
> (Smithers and Walker, 2000: 841)

This demands improved HRP at two levels: HR-business planning integration and cultural management. Chinowsky and Meredith (2000) further focused the importance of strategic planning at organisational level, where Kang *et al.* (2001) concentrated on the unit of production within the industry, the project, in developing a computerised model for optimal schedule planning. Shi and Halpin (2003) attempted to integrate the two by utilising a construction enterprise resource planning system. All usefully recognised the specific constraints the nature of construction work places on the planning process; however, failed to take into account the HRP consequences. Loosemore *et al.* (2003: 84–89) noted the complications the competitive tendering process and cyclical nature of construction markets bring to effective HRP. In particular, they highlight the dilemma of laying staff off in response to a downturn versus problems in recruitment during upturn. They reason that this influences organisations to hold an apparently inefficient surplus of labour during recessionary periods. Indeed, *flexibility* is one of the key drives/barriers for effective HRP in construction organisations.

Flexibility

Flexibility is a key concern for modern people management practice. It is a necessary requirement for all parties entering the employment contract, be that the employer, employee or a third party representing either one of these. Taylor (2002b: 403) highlights two important reasons:

- An organisation that is flexible is able to deploy its people and make use of their talents more effectively and efficiently than one that is not.
- The more flexible an organisation becomes, the better able it is to respond to and embrace change.

Atkinson's (1981) model 'flexible firm' (Figure 3.3) is one of the most influential and widely debated illustrations of flexibility within organisations, and has been variously applied to construction organisations (e.g. Loosemore *et al.*, 2003; Langford *et al.*, 1995).

By the term 'flexible firm', Atkinson referred to an organisation that is competitive in the modern business environment. He intended the model as an illustration of flexibility as a form of employment strategy.

The model (Figure 3.3) proposes a break up of the traditional hierarchical structure of the organisation and suggests that radically different employment policies can be pursued for the different groups of employees (Atkinson, 1984). The different groups of employees include two types within the organisation's internal labour market: the core group and first peripheral group. External sources of labour are provided by the second peripheral group and sub-contracting, outsourcing, agency temporaries and self-employed. The internal labour markets exercise mainly functional and numerical flexibility. The core group consists of stable, key staff which conducts the organisation's central activities. At this core, only tasks and responsibilities change (functional flexibility) and so the employees are

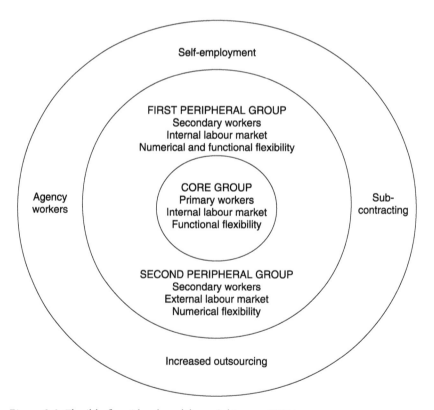

Figure 3.3 Flexible firm (developed from Atkinson, 1981).

protected from the fluctuations in the economic markets. The first peripheral group is less stable but the employees are often full-time as those within the core group. This group however have fewer opportunities for career development or long-term job security; numerical flexibility is often sought in response to market fluctuations.

The second peripheral group and other external sources of labour (subcontracting, outsourcing, agency temporaries and self-employed) seek to 'maximise [numerical] flexibility while minimising the organisation's commitment to the worker' (Atkinson, 1984: 29). Jobs within this category are often highly specialised and therefore carried out by specialist contractors (outsourcing) or mundane, such as office cleaning (part-time).

Although Atkinson's model has been extremely influential in initiating wide reaching debate internationally (OECD, 1986, 1989; Brewster, 1998) it has also received heavy criticism on its limited utility and lack of theoretical robustness (Pollert, 1988). In addition, the model is rather restricted in its sole focus on flexibility as a strategy employed by the organisation to achieve competitive advantage.

As modern business organisations now commonly adopt multiple and parallel forms of flexibility in response to pressures (such as the changing nature of work, increased use of technology and the need for improved operations), the traditional forms of flexibility incorporated within the flexible firm model (functional, financial and time flexibility) have been updated by a more complex set of flexibilities which more closely reflect the ways in which such businesses operate (Sparrow and Marchington, 1998: 18–19):

- *Numerical flexibility* – does the job need to be one within the internal labour market or can it be sufficiently controlled through outsourcing or other type of peripheral form of employment?
- *Functional flexibility* – what are the roles and competencies deemed appropriate for the job, does the job need to be staffed by a multi-skilled individual, are there core competencies that must be delivered, or important business process skills that need to be acquired?
- *Financial flexibility* – what is the best balance between the type and nature of reward and the delivery of performance?
- *Temporal flexibility* – what time patterns should the job be fitted into, will the employees be able to deliver the highest level of customer service and business performance through these time patterns?
- *Geographical flexibility* – does the job need to be carried out in specific locations or is there latitude for teleworking/virtual teams?
- *Organisational flexibility* – does the organisation operate as adhocracy/ a loose network of suppliers, purchasers and providers/temporary alliance/joint venture?
- Cognitive flexibility – does the job require people with a particular type of psychological contract, what sorts of strategic and cognitive assumptions cannot be tolerated?

These broaden the focus of flexibility from the organisational strategy for competitive advantage to include the employee perspective and accommodation of their needs and preferences. Indeed, flexibility has received increasing attention within the strategic HRM literature as an organisational tool for retaining staff. Thus, the Sparrow and Marchington's set of flexibilities provide a more accurate reflection of the use of flexibility within modern organisations than that introduced by Atkinson's flexible firm.

Volberda (1997) expands on Sparrow and Marchington's view of flexibility highlighting different types of internal and external *organisational flexibility*. He categorises these under 'routine', 'adaptive' and 'strategic' levels, as shown in Table 3.4. This usefully provides the wider context for discussing flexibility, simultaneously incorporating elements of strategic HRM, such as use of temporary labour (external operational flexibility) and creating

Table 3.4 Examples of internal and external flexibility (Volberda, 1997: 171, with permission from Elsevier)

Capacity for manoeuvre	Type of flexibility	
	Internal	*External*
Routine	*Internal operational flexibility*	*External operational flexibility*
	Variation of production volume	Use of temporary labour
	Building up inventories	Multisourcing
	Use of crash teams	Reserving capacity with suppliers
Adaptive	*Internal structural flexibility*	*External structural flexibility*
	Creating multifunctional teams	Purchasing components from suppliers with a short delivery time (JIT)
	Changing managerial roles	Purchasing sub-assemblies from suppliers (co-makership)
	Alterations in control systems	Developing sub-components together with suppliers (co-design)
Strategic	*Internal strategic flexibility*	*External strategic flexibility*
	Dismantling current strategy	Creating new product market combinations
	Applying new technologies	Using market power to deter entry and control competitors
	Fundamentally renewing products	Engaging in political activities to counteract trade regulations

multifunctional teams and changing managerial roles (internal structural flexibility), into the approach.

Flexibility is highly important to construction organisations. Risk and uncertainty is an every day feature of project and organisational management within the industry. Thus, an accommodating approach to (especially unexpected) change is required. This applies equally to the production related issues as well as human resources. Walker and Loosemore (2003) present an effective *solution-building ethos: best-for-project culture*. This focuses on detecting emerging problems early and dealing with them quickly at a project level. Such an ethos also inevitably leads to short-term thinking and a lack of strategic planning within such firms. This runs counter to the longer range focus needed for building capabilities under the resource-based view of the firm described earlier.

At company level, Langford *et al.* (1995: 55), Druker and White (1995: 88), Loosemore *et al.* (2003: 54–55) and most recently Druker (2007) noted the relevance of Atkinson's flexible firm model to the way construction workforce is organised. Construction organisations' project and operational senior managers easily fit into the core group and the use of the first peripheral group's numerical flexibility allows 'untroubled and speedy adjustment to changes and uncertainty in the construction services market' (Langford *et al.*, 1995: 55). Most importantly however, the external sources of labour such as sub-contractors, agency temporaries and self-employed, are very common in construction. This necessitates key management skills to co-ordinate the sub-contract activity (Druker, 2007).

However, the increased need for organisational flexibility has contributed to a growing sense of job insecurity among employees as companies have undergone radical restructuring, delayering or downsizing (Sparrow and Marchington, 1998: 16). These changes in organisational structures have led to flatter managerial hierarchies with few opportunities for vertical progression. Jobs have been rationalised, communication links streamlined and functional barriers brought down. Thus, while flexibility may be an effective method of coping with fluctuating markets and dynamic staffing requirements, there remain many advantages to maintaining a large primary labour market within construction organisations. Indeed, some construction organisations are now recognising these dangers and increasing, once again, their numbers of core employees.

Work–life balance

Another priority in HRP within the modern construction organisation is to take account of work–life balance priorities. This issue has become increasingly important over the last few years, not only because of changes in societal values. The escalating pressures for high performance have resulted in many employees to work increasingly long hours, which often conflict with their family or other outside work commitments. This, together with the intensification of work, has often led to stress-related problems. In order to combat the absence resulting from stress-related problems and shortages in availability of suitably qualified candidates for recruitment and selection, many organisations have recognised their responsibility to assist their employees in achieving work–life balance. Family-friendly policies and flexible forms of working are common initiatives organisations have adopted in attempts to accommodate the employee needs and thereby facilitate staff retention. These include assistance with childcare, special leave arrangements (maternity/paternity/parental/emergency leave), home- or teleworking, part-time and/or term-time working, annual hours and flexitime. Although powerful retention tools, such arrangements have implications for staff deployment, a central employee resourcing activity.

Long working hours are commonly seen as an inherent characteristic of construction work (Lingard and Sublet, 2002: 507; Strategic Forum for Construction, 2002: 31). Tight project deadlines and seasonal changes in weather conditions require significant flexibility on the employees' availability for work. This results in many juggling to balance their work and family commitments since few construction organisations offer their employees assistance on work–life balance issues. Dainty and Lingard (2006) outline the factors which render construction such a problematic sector from a work–life balance perspective. They suggest that travel, long hours and high levels of stress or endemic within the sector, and highlight that employers place importance on the flexibility of their employees, who are expected to balance their work and family commitments, often with little organisational assistance (see Foley, 1987). Indeed, long hours create a high-stress environment, which can even result in suicide. This underscores the need for construction organisations to recognise the demands of family and other personal responsibilities that their employees face, and genuinely attempt to accommodate these (Lingard and Sublet, 2002; Loosemore *et al.*, 2003).

Recruitment and selection

The recruitment and selection processes are concerned with identifying, attracting and choosing suitable people to meet organisational needs. While recruitment and selection activities are inherently integrated, some distinguish *recruitment* as the 'searching for and obtaining potential job candidates in sufficient numbers and quality' so that an organisation can select the most appropriate people to fill the vacancy (Dowling and Schuler, 1990). Thus, *selection* becomes an activity concerned with predicting which candidates will make most appropriate contribution to the organisation.

The reconciliation of the HRP outcomes with the shorter term operational conditions of the business indicates the levels of recruitment required. Larraine and Cornelius (2001) highlight the importance of the following:

- analysing the organisation's long-term resource requirements;
- clear advertising of the vacancies via appropriate media thereby ensuring the widest possible pool of suitable candidates is attracted and much desired choice in the selection process achieved;
- determining appropriate reward linking the process with other HRM strategies;
- the measurement, review and evaluation of the selected candidates' performance being fed into the organisation's performance management systems.

By emphasising the continuity of the process and the links with other HRM systems Larraine and Cornelius' approach reveals the vital importance of effective recruitment and selection process; ensuring an appropriate supply of skilled staff that can positively contribute towards the achievement of business objectives. Analytically, this type of recruitment and selection is often referred to as a 'systematic approach'.

De Feis (1987) recognised the crucial role of employee recruitment and selection in effective people management in construction organisations. The current and estimated future skills shortages in the industry have recently highlighted the need for increased recruitment efforts. Fellows *et al.* (2002: 119–122) note this being important at three levels: the industry, trade/profession and company.

Naturally, the recruitment of appropriately skilled candidates to the industry and the trades/professions within it are seen as having a crucial impact to attracting a suitable pool of potential candidates at the company level. National bodies, such as the Construction Industry Council (CIC) and the Sector Skills Council (ConstructionSkills) actively promote the image of the industry. At the company level, this vital task is generally managed by line managers rather than specialist HR personnel. Informal practices and personal introductions and contacts are common and important source of recruitment at all levels. Selection methods are usually restricted to interviews and assessment centres, although the strategic HRM literature suggests that a much wider range of techniques may be beneficial. The Housing Forum (2002) put forward a range of initiatives that could enhance the industry's ability to recruit. These include promoting the industry within schools and higher education institutions and diversifying organisational recruitment and selection procedures to target wider audiences, such as women, ethnic minorities and the unemployed. The poor image of the industry is seen as detrimental to recruitment, and as leading to the exclusion of potential entrants such as women and ethnic minorities (Strategic Forum for Construction, 2002; CITB, 2003; Loosemore *et al.*, 2003).

Team deployment

The staffing function is also increasingly concerned with team formation and development, and the deconstruction and redeployment of teams. The fundamental requirement of effective team deployment is to select team members carefully on the basis of their skills and competencies. This is crucial; enforced changes in key project personnel are highly disruptive to project performance. Many typologies and approaches to achieving effective team composition exist. For example, Belbin's (1991) team role model and Margerison and McCann's (1991) team management wheel can aid the selection of individuals who together form a balanced and complementary workgroup. However, effective team performance also relies on members' abilities to successfully integrate their individual actions. In addition, the organisational climate affects the success of a project team. Characteristics, such as freedom of expression, participation in the definition of goals (employee involvement) and innovation all positively impact project outcomes. Ideally, 'a low threat, secure and stable environment in which individual contribution is maximised within a distinctive team culture offers the optimum environment for successful project outcomes' (Gray, 2001). In reality however, this kind of climate rarely exists within the modern business environment and so effective leadership is vital in defining team direction and in ensuring their optimum performance. Flexibility is a key requirement to which the team deployment activities must respond. As discussed earlier (under *human resource planning*), the use of multiple and parallel flexibilities help ensure that an organisation (i) is able to deploy its people and make use of their talents effectively and efficiently, (ii) is able to respond to and embrace change, and (iii) is able (and willing) to accommodate the employee needs and preferences with a view of retaining staff. Sole focus on organisational competitive advantage potentially undermines the psychological contract (see *change management* hereafter) and leads to an exit from the organisation.

Effective teamwork is described as the *key to competitive advantage* in the construction industry (Druker *et al.*, 1996). It plays a central role in the industry's operations due to the project-based nature of work. Yankov and Kleiner (2001) also identify it as a powerful motivator: team belonging fosters a feeling of participation in a group. This supports the achievement of the project/organisational goals leading to improved productivity. Good team relationships are essential for achieving such success. Project leaders play a key role in building and maintaining productive team environment. Odusami (2002) found leadership, motivation, communication and effective decision-making as the central qualities of a resourceful project leader. This supported his literature search findings, which repeated 'soft' people skills throughout as crucial elements of successful project management (see Odusami, 2002: 63, Table 2). Spatz (1999) further pointed out that outstanding leaders assess their own abilities to lead. This creates a powerful condition for successful leadership: trust.

The ultimate goal of a teamworking environment is a co-operative culture where managers and employees grow and prosper together. Team effectiveness coaching can help create such a culture by improving team interaction and thereby increasing its technical qualities (Goldberg, 2003). This follows the argument that the intangibles of human interaction frequently separate average performance from outstanding execution (Spatz, 1999, 2000; Goldberg, 2003).

Despite the recognised importance of teamwork for construction organisations, relatively little attention has been paid to the actual process of building or composing effective teams. Walker's (1996b) work on winning construction teams in terms of project time performance in Australia provides some indicators to successful team deployment. In short, Walker found that the responsibility for good construction time performance lies firmly with the management team. However, his work placed considerable emphasis on the management team's ability to influence and manage the construction *process*, rather than team interaction as suggested by Spatz (1999, 2002) and Goldberg (2003). Ogunlana *et al.* (1999, 2002) found that, at least in the Thai context, the project manager's ability to meet the external clients' needs was considered as the most important project management quality in deploying leaders to projects. The project manager's technical experience, relationships (with clients, project team and management) and management abilities (planning, controlling, directing, organising and staffing) were found the top three factors that influenced team leader selection. Leadership capabilities were ranked sixth out of 11 categories. While these reports contrast with Odusami's findings earlier, Ogunlana *et al.* (1999, 2002) recognise the limited scope of their study (i.e. to Thailand) and thus recommend for their findings to be viewed in the context of the societal culture. This could explain the differences between Ogunlana *et al.*'s (1999, 2002) and Odusami's (2002) work.

In conclusion, clearly both, the 'hard' (technical/process) *and* 'soft' (leadership/team interaction) factors contribute to the success of the project, and so both should be included as integral aspects of team deployment decision-making.

Exit

The monitoring and management of exit from the organisation is best carried out on an on-going basis and its outcomes applied into the organisational learning processes as they emerge (Huxtable and Cheddie, 2002). There should be a strong link with human resource planning. The management of involuntary forms of exit, such as retirement, redundancies and dismissals, should adhere to legal and procedural guidelines.

In the past, managing retirement was a straightforward activity. At a nationally set retirement age, all employees could be dismissed and those who wished to remain in the workforce worked within informal arrangements. The realisation of 'ageing population' and the introduction of legislation that protects employees from discrimination on the grounds of their age (in

December 2006) changed this. Employers can now set any retirement age they want and dismiss all employees who reach that age. At the same time, employees who remain at work beyond their retirement age cannot be dismissed on the grounds of their age alone. There are many reasons why employees choose to work beyond their retirement age. Very often financial reasons necessitate this, but sometimes it is down to personal motivation use skills and experience and to provide meaningful contribution to society (CIPD, 2008). The statistics on ageing show that not only is there a world-wide general trend that life expectancy is increasing, but that people are remaining healthy and so are employable for longer. There are strong argu-ments for encouraging people to work longer, principally to retain their skills and experience, but on a more general level to ease the burden of supporting a retired population on a reducing proportion of the working population. These general arguments apply more to managerial and office staff in the construction industry than site operatives because of the physical demands of the work (for further information on ageing, see www.sparc.ac.uk).

Redundancy is defined as 'a situation in which, for *economic* reasons, there is no longer a need for the job in question to be carried out in the place where it is currently carried out' (Taylor, 2005: 360). The reduction in the need for employees in general or at a particular location can be in response to reduction in volume of work or type of work. For example, an introduc-tion of new technology often means that fewer low-skilled workers are required. Sometimes the need for redundancies arises out of a strategic deci-sion to change organisational structure. This may result in the reduction in layers of management, for instance. However, commonly restructuring is less about economics and more about managerial choice to downsize or 'rightsize' an organisation.

Whatever the reason for redundancies, it is one of the most traumatic events an employee (and an organisation) may experience. Announcement of redundancies will invariably have a destabilising impact on morale, motiv-ation and productivity within the whole workforce (Redman and Wilkinson, 2006). Thus, redundancies should be avoided where possible. In the long-term, this can be done through effective human resource planning and flexible working arrangements. In the short-term, initiatives such as wage reductions, part-time work, career breaks, redeployment, reductions in pay costs (such as cut in overtime and bonuses) and non-pay costs (such as reduction in the office area/space used or hiring less expensive premises), recruitment freeze, call for volunteers and offering early retirement can be considered.

Given the susceptibility of construction to be affected by fluctuations in wider economic growth (i.e. this reduces capital expenditure which in turn, reduces spend on buildings infrastructure), it is inevitable that in times of recession some construction firms are faced with an over-capacity of workers. Whereas outsourcing and flexible employment structures have mitigated this to some extent (see aforementioned), it is essential that where they do occur, the employers handle redundancy situations effectively and sensitively. The CIPD (2008) emphasises that the negative effects can be reduced by sensitive handling of redundant employees and those that remain in the organisation. There are four areas that must be planned and executed vigilantly: selection for redundancy, consultation and communication, employee support and managing the survivors (see CIPD, 2008, for good practice guidelines). Confusion sometimes arises because 'making someone redundant' is often used as an euphemism for saying an employee is being dismissed for some reason other than redundancy.

Dismissal requires sensitive handling. Legally, dismissal occurs when an employer terminates the contract, either with or without giving notice; a fixed term contract ends and is not renewed; or an employee leaves, with or without giving notice, in circumstances in which they are entitled to do so because of the employer's conduct (Soret, 2007). Where a dismissal is considered 'fair' the employer has a good reason for the dismissal and has acted 'reasonably' in carrying it out. A 'wrongful' dismissal refers to a breach of contract. Some circumstances warrant an automatically 'unfair' dismissal, for example where the reason for dismissal relates to the following:

- trade union membership or activities
- pregnancy or childbirth
- taking maternity, adoption, paternity or parental leave
- asserting a statutory right
- claiming the National Minimum Wage
- asserting rights under the Working Time Regulations
- transfer-related reason under the Transfer of Undertakings (Protection of Employment) Regulations 2006
- exercising the right to request flexible working.

Employment equality legislation includes further provisions to consider when deciding 'fair' criteria for dismissal.

While many legal and good practice guidelines are necessary for managing the involuntary forms of exit; absence, employee turnover and other voluntary means of exit from the organisation may be managed usefully through performance and career management processes.

Performance management and career development

Performance and career management mechanisms focus on maximising individual, team and organisational performance while facilitating employees' career development. Performance management systems, particularly those aimed at evaluating team performance via a composite of qualitative and quantitative measures, can also help assess the complex sum of variables that contribute to effective team/project performance. These are often referred to as 360 degree feedback mechanisms. Skilfully operated systems provide a useful tool for managing the balance between the competing organisational, project and individual employee priorities, needs and preferences. However, there are several difficulties in implementing particularly more complex systems. The instruments may become large and cumbersome as feedback is sought from many aspects of the job. Sometimes the systems may have conflicting purposes, for example to provide the basis for reward (increases) and outline development needs. In all cases, employee involvement (see *change management* hereafter) is a key to the success of performance and career management initiatives and systems. Unless employees feel an integral part of the process, they are unlikely to buy into the long-term commitment required to achieve desired results.

The construction industry's performance improvement agenda has challenged organisations to strengthen their business and management practices in order to improve the overall efficiency, quality, productivity and cost effectiveness of the industry. Kagioglou *et al.* (2001) investigated the literature on performance management and measurement at an organisational level with a view of transferring best practice into construction. They identified that traditionally, performance measurement within the industry was approached in relation to the product as a facility and in relation to the product as a process. 'Hard' client objectives, such as cost, time and quality, were identified as the dominant factors in assessing the success/failure of construction projects. It was argued that in isolation these do not provide a balanced view of the project's performance. 'Softer' factors, such as people involved, impressions of harmony, goodwill and trust, need to be included in the equation.

Moore and Dainty (1999) recognise the relationship between effective team integration and performance: engendering a single focus and culture of co-operation within a team should outweigh the additional costs of bringing the team members together at the outset of a project and maintaining close physical proximity through an improved response to unexpected change events, better project performance and hence, enhanced client satisfaction. This highlights the importance of considering the team and related team deployment issues in assessing project/organisational performance. Belout (1998) and Nicolini (2002) further emphasised the significance of human resource and social and cultural factors on project success. Nicolini (2002) termed a well-integrated team within which good and open communications facilitated effective collaboration, psychological safety and care and the achievement of shared goals 'project chemistry'. His findings suggested that good 'project chemistry' binds the team together and thus enhances the quality of the final product by harvesting expertise and creativity from all team members, reduces time through early detection of potential problems and improved problem-solving process, constitutes a motivator and helps people work in the same direction, and allows for more productive use of resources and less defensive bureaucracy.

At an individual (employee) level, Glad (1994) identified the act of balancing the needs of employees with the needs of the business the most challenging performance issue. According to her findings many managers fear that they will either cave in to employee demands at the expense of better business judgement, or act too mechanistically in serving the business thereby creating a people problem. Confronting performance problems was identified as the most difficult arena. In response, Glad (1994) suggested a four-step programme:

1   Managers must realise that staff turnover is costly and consider the employee's present and potential value to the company.
2   The problem, or a 'performance gap', must be defined in a measurable way.
3   The nature of the inferior performance must be determined: is it 'can't do' or a 'won't do' problem.
4   A realistic action plan for improvement, which involves both the manager and employee must be set and implemented.

Shah and Murphy's (1995) study in engineering suggested the performance appraisal as an effective tool for managing and improving employee (and organisational) performance. They found an increasing use of the tool within civil engineering organisations, with the benefits of increased motivation and productivity. Nevertheless, areas for improvement were noted: consistency in the process and timing, training, follow-up, commitment from the top management and open discussions. Druker and White (1996: 122) identify the disparate location of construction projects as the main problem to effective performance management within construction organisations. Managers tend not to have close frequent contact with their employees and thus see little of their day-to-day performance. In line with Shah and Murphy's (1995) findings, they note that although appraisal systems exist within most organisations in the industry, they are often ad hoc and informal.

Nesan and Holt (1999: 168) suggest 'group assessment' as an effective way forward for performance measurement. This is similar to the 360 degree feedback mechanism in that it is based on performance data from a diverse range of sources (self, team, department, financial, senior management) and senior management takes the responsibility for benchmarking company-level performance with identified business competitors as well as publishing the outcomes back to the employees and external customers/suppliers highlighting both successes and failures. However, Loosemore *et al.* (2003: 100) criticise this technique on the grounds that the very breadth of such a system makes it costly and time consuming to operate and requires a considerable staff resource. Besides, unless an open culture of trust and honesty is firmly embedded within an organisation, 360 degree feedback or group assessment is likely to produce corrupt results. Particularly the appraisee's subordinates may fear retribution and therefore provide favourable feedback regardless of their true opinions. This ties back to the areas for improvement identified by Shah and Murphy (1995): consistency in the process and timing, training, follow-up, commitment from the top management and open discussions are a must if appraisals are to succeed within the industry.

Within the UK, there has been an emphasis on measuring performance against competence standards. This emphasis is rooted within the Employment Department's Standards programme which defines competence as 'a description of something which a person who works in a given occupational area should be able to do, it is a description of an action, behaviour or outcome which a person should be able to demonstrate' (Training Agency, 1988). The approach has been criticised for the inappropriate and inflexible standards that the functional analysis approach promotes (Cole, 2002: 368). Such a model cannot take account of the dynamic context in which managerial behaviours are enacted. Indeed, effective managers do not simply apply required actions, but reflect on their actions, experiment, and in doing so, learn and develop themselves (Kolb and Fry, 1984). Competency differs in that it refers to the underlying characteristics of a person which result in effective actions and hence, superior performance in an occupational role (Boyatzis, 1982). Thus, competency models can help to refocus employees on what it takes to succeed (Brophy and Kiely, 2002).

It is easier to apply the concept of competence to employee selection and management development, where underlying personal characteristics are the key determinant of success, than it is to the assessment of job performance (Elkin, 1990). This is because focusing on job performance results in long detailed lists of job task micro-competency statements which can be difficult to measure performance against. Thus, competency profiles arguably offer an improved benchmark against which managers' performance can be assessed. However, failure to link competencies to appraisal and reward seriously delimits their value in underpinning organisational growth and development (see Abraham *et al.*, 2001). For an example of a competency-based performance framework and how it can be applied (see Dainty *et al.*, 2004b).

Careers

Over the past decade, environmental pressures, the individualisation of the employment contract, changing nature of strategic HRM and increased demand for flexibility have contributed towards fundamental changes in careers (Arthur and Rousseau, 1996; Thite, 2001; Baruch, 2003). The traditional hierarchical structures have been replaced by more open systems within which an individual is required to navigate his/her way with only minimal support from the organisation. Thite (2001: 313) usefully summarises this (see Table 3.5).

Traditionally the organisational personnel policies and structures clearly directed their employees through almost pre-defined careers. Under the contemporary framework many organisations have responded to the challenging context by abandoning practically all responsibility for career management. According to Schein (1996), this is due to the fact that both

Table 3.5 Traditional and contemporary framework of career management (Thite, 2001: 313, with permission from Elsevier)

	Traditional framework	Contemporary framework
Environmental context	Production driven Protected markets Stable technology Familiarity with domestic political, legal and cultural framework	Era of discontinuity and hyper competition at a global level Service driven Technology intensive Global markets with unpredictable economic, political and cultural scenarios
Organisational response	Growth at any cost business strategy Mechanistic, product, functional, divisional structures Hierarchical, multiple management levels Supervisor-based, time-bound promotions Command and control management style Responsible for individual career planning and development Unidimensional career movements (ladder)	Knowledge and information technology-driven learning organisation Strategic collaboration with competitors Network, cellular structures Small component of core employees and big component of part-time, casual and contract staff Empowerment of people 360-degree feedback Competency based outsourcing Self-directed teams Delayering Multidimensional career movements (jungle gym)
Individual response	Loyalty to organisation in return for lifelong and steady growing employment Minimal responsibility for career management Emphasis on specialisation of skills Collective bargaining of employment issues	Diminishing loyalty for organisation Focus on employability rather than job Portfolio of jobs and skills Increasing emphasis on life-style issues Acceptance of near-total responsibility for career management Life-long learning

the organisation and the individual are gradually adjusting to the notion that they have to look out for themselves, meaning that organisations will become more paternalistic and individuals more self-reliant. Others argue, however, that careers are still to an extent 'property' of the organisation and hence should be managed by them. It is recognised that the traditional bureaucratic framework is no longer feasible. Instead, a normative model for organisational career management may be more appropriate in the context of the modern business environment. Baruch (2003) identifies this type of an approach to incorporate a portfolio of organisational career practices and analytical dimensions. The organisational practices are as follows:

- *Posting (advertising) internal job openings.*
- *Formal education/tuition reimbursement.*
- *Counselling by manager/HR* (two-way communication between manager/HR representative and employee on career issues).
- *Lateral moves/job rotations* (job transitions/moves at the same hierarchy level within the organisation aimed at creating cross-functional experience, particularly relevant where fewer hierarchy levels exist and horizontal communication is a key to success).
- *Pre-retirement programmes* (directed at a target population approaching retirement, aims to ease the transition from full working life to retirement).
- *Succession planning, formal mentoring and common career paths.*
- *Dual ladder* (a parallel hierarchy created for professional/technical staff which enables them upward mobility and recognition without a move to a managerial role particularly suitable for professionals without managerial skills or no intention of becoming managers).
- *Career booklets/pamphlets and written individual career plans.*
- *Assessment centres* (a reliable and valid tool used to evaluate people in an extended rigorous work sample process, usually specifically designed for evaluating the potential of present or future managers) and *development centres* (directed towards general development and enhancement of particularly managers, preparing them for future roles).
- *Use of performance appraisal or 360°* feedback for career planning (performance appraisal assesses and measures employee performance against agreed objectives, 360° feedback is a multi-rater appraisal mechanism which incorporates the views of peers, subordinates, internal and external customers and the manager).
- *Induction/orientation programmes* ('socialising' new employees into the organisation) and *career workshops* (short-term workshops focusing on specific aspects of career management, such as identifying future opportunities or improving employability, with the aim of providing managers (and employees) the relevant knowledge, skills and experience).
- *Special attention* (e.g. high-flyers, dual-careers couples) and *equal opportunities/managing diversity population* (e.g. age, gender, minorities – practices not necessarily concerned with discrimination, but providing support).
- *Creating (and maintaining) balanced psychological contracts* (from provision of realistic job previews, through fair and open communication, to open discussion on exit).

- *Secondments* (temporary assignment to another area within the organisation, or externally, aimed at generating different/holistic perspective of the organisation).

The analytical dimensions include strategic orientation, developmental focus, decision-making and innovative approach. The strategic orientation means that strategic HRM (and its individual components) are managed as an integrated comprehensive strategy, which is applied within the organisation's overall strategic management. Developmental focus promotes investment in the development of people, the organisation's core asset and source for competitive advantage, with the presumed reward of improved performance, effectiveness and efficiency. Decision-making seeks to identify and align the appropriate strategic HRM practices with the organisational strategic decision-making. The innovative approach incorporates novel ideas and concepts, such as the intelligent career (basic investment and development of know-how), the boundaryless career (managing careers outside the organisational boundaries), the post-corporate career (the need to rethink the type of relationship the organisation has with its employees, that is it is based on a traditional contract of employment/sub-contracted/etc.), the protean career (major role of career management on the individual) and career resilience (the need for organisations to 'educate' their employees and incorporate them into the realm of instability where employability rather than long-term employment is the norm). This type of approach formulates a broad comprehensive career system, which can help balance the responsibility of career management between the individual employee and organisation and support both sides in this crucial area (Baruch, 2003: 244).

Along with developments in other sectors changes in the business environment have initiated fundamental transformation in construction careers over the recent years. Much in line with Thite's (2001) contemporary framework (Table 3.5), it is now the employees who control the direction of their own careers and hence focus their efforts on personal and professional growth (Schirmer, 1994; Weddle, 1998). The growth is achieved via continuous development and inter-organisational job moves, for which the recent increase in business activity within the industry has created fresh opportunities. As long as the people are maintaining an overview of the job market situation and aligning their capabilities with those required by the organisations, they can manage their careers instead of reacting to the environment.

However, a return to Baruch's (2003) argument of careers still to an extent being the property of the organisation and therefore needing to be managed by them provides an alternative perspective in contrast of this highly individualistic view. Indeed, Dainty *et al.* (1999, 2000a, 2000b) suggest that it is essential for construction organisations to address the current career/developmental barriers that hold back the progress of female professionals within the sector if they are to tap into the best talent available within the society.

Career management/development opportunities tend to influence other areas of strategic HRM, for example reducing or increasing the pool of potential candidates for recruitment and selection. Accordingly, it is crucial for the success of integrated strategic HRM within the industry that construction organisations provide their employees with a flexible range of career structures and varied opportunities for personal and professional development (Schirmer, 1994; Fellows *et al.*, 2002). This implies the need for organisations to resume their part of the career management responsibility, despite the recent trend to transfer the responsibilities to their employees, and work in partnership with their employees. A partnership approach to career management and development would also increase the level of employee involvement in the industry. The normative model proposed by Baruch (2003), within which both the employee and manager actively participate in decision-making, may provide an appropriate solution for managing modern careers in construction too.

HR administration

HR administration focuses on the collection, storage and use of employee and organisational data in support of the HR function. Contemporary Human Resource Information System (HRIS) solutions provide sophisticated instruments to aid this process. The most advanced web-enabled software interfaces with other administrative programmes and include 'self-service' capabilities whereby individual employees update their own records which can subsequently be used in the strategic HRP activities. This makes process integration easier, reduces managers' administrative workload and encourages employee involvement, among numerous other benefits (Raidén *et al.*, 2008).

Human resource information systems

Information Technology is often seen as an effective stimulus for achieving transformational change. 'Human resource information systems' is the term used to refer to a particular type of software that is aimed at supporting the strategic HRM function within organisations. Broderick and Boudreau (1992: 17) define HRISs as 'composite of databases, computer applications and hardware and software that are used to collect/record, store, manage, deliver, present and manipulate data for human resources'. In short, HRISs provide an electronic database for the storage and retrieval of employee data that offers the potential for flexible and imaginative use of this data (Tansley *et al.*, 2001: 354). Four main types of HRIS applications are given as follows (CIPD 2004: 6):

1   A single HRIS that covers several HR functions which are integrated within the system itself but not with any other system within the wider

organisation. This is the most common type of HRIS in use (59% of the CIPD 2004 survey respondents indicated using such system).

2   A single HRIS which covers several HR functions that are integrated within the system itself and with other IT systems within the wider organisation (21% of the respondents to the CIPD 2004 survey had this type of system in place).

3   Multiple systems with two or more stand-alone HRIS packages that cover different HR functions, but are not integrated with each other or other organisational IT systems (14% of the respondents to the CIPD 2004 survey had this type of system in place).

4   Multiple systems with two or more stand-alone HRIS packages that cover different HR functions and are integrated with other IT systems within the wider organisation. (This approach is relatively rare. Only 6% of the respondents to the CIPD 2004 survey indicated using such systems.)

These integrated suites of HRIS work packages are now widely available, both as the stand-alone solutions or global, enterprise-wide information systems that support many functional and operational pillars of an organisation. While some of the HRISs available are simply sophisticated database applications, some incorporate expert systems/artificial intelligence into the system thereby increasing the system's learning capabilities. An expert system is 'a computer program that represents and reasons with knowledge of some specialist subject with a view to solving problems or giving advice' (Jackson, 1999: 2). It solves problems by heuristic or approximate methods, which do not require perfect data. Thus, expert systems have the benefit of being able to propose solutions with varying degrees of certainty. Another significant benefit of an expert system is that its workings are transparent: the system is capable of explaining and justifying solutions or recommendations in order to enable the user to ascertain that its suggestion is sound.

Accordingly, more recent work proposes that HRISs seek to go beyond the 'electronic filing cabinet', vertically and horizontally integrating HRM practices and processes in order to enable and transform strategy-making and thus add value to the organisation (see Table 3.6). This step change is described by Fletcher (2005) as an evolution of E-HR through three phases: the efficiency and control phase; the enabling insight or partnership phase; and the creating value or player phase.

This table highlights only few of the key functionalities of HRISs and their associated benefits. Many writers and professional practitioners have recognised the substantial benefits that HRISs can bring to the efficient management of the HR function (see Broderick and Boudreau, 1992; Ettorre, 1993; Greenlaw and Valonis, 1994; Kossek *et al.*, 1994; Hosie, 1995; Kinnie and Arthurs, 1996; Edward, 1997; Eddy *et al.*, 1999; Ball, 2002; CIPD, 2005). The systems have particular capabilities for managing staffing, learning and development, performance, reward and HR administration. They can help

Table 3.6 Uses and benefits of HRISs (after Tansley *et al.*, 2001)

Automate	Inform
Use of the system	
Electronic filing cabinet	Sophisticated database/expert system
Enables storing and analysis of employee data	Enables managers to act on HR information
Support more effectively direct control – employee activities and productivity transparent to managers	Assumes a philosophy that the system itself and appropriate managers can make decisions – provides access to comprehensive range of information
Facilitate close supervision and monitoring	Facilitates empowerment and indirect control
HR access	Employee self-service
Benefits	
Task mechanisation – can save mental and/ or physical labour in data management	Can transform HR practices
Process automation – can enable greater efficiency of HR practices	Can enable managers to integrate their business objectives with HRM priorities
Cost reduction (in reduced overheads)	Cost reduction (in reduced overheads)
Improved HR service: faster service, improved quality and consistency of information	Availability and accessibility of wide range of information

HR professionals to improve productivity, control employee benefits, streamline compliance with HR legislation, manage the payroll function, and lower the costs of employee resourcing. As outlined in Table 3.6, in essence they automate daily administrative HR tasks, integrate cross-departmental activities and ensure the accuracy and consistency of employee records.

The recent developments have led to HRISs having the potential to hold comprehensive databases of employee skills and qualities, including their future aspirations, and produce complex reports mapping the employee abilities and preferences against forthcoming vacancies/projects. The latest generation of web-enabled HRISs now also allow employees to update their own personnel records, submit timesheet data, review benefits, request holidays and enrol on training courses. This integration of so many key strategic HRM activities leaves HR professionals and line managers more time to focus on strategic activities, and provides information for them to be able to turn their employee assets to a source of competitive advantage. Thus, HRISs are revolutionising the strategic HRM function by providing up-to-date information, services to employees, return on investment, and strategic analysis and partnership (Greenlaw and Valonis, 1994; Miller, 1998; CIPD, 2004). Tansley *et al.* (2001: 364) conclude '... introduction of the HR system could potentially provide the stimulus to actually effect the required change in employee management practices . . .'.

Nationally there has been a steady increase in the number of organisations that have a HRIS and now three quarters (77%) of organisations report using such systems (CIPD, 2005). The main use of a HRIS is to 'automate' administrative tasks such as absence management (90%), reward management (75%), and monitoring training and development (75%) (CIPD, 2004). Less than a third of organisations use their HRISs for strategic purposes: HR planning (29%); HR strategy (18%); knowledge management (25%) (CIPD 2004; CIPD 2005). This implies that the majority of organisations have not yet moved beyond the efficiency and control phase (Fletcher, 2005: 2). There may be a number of reasons for this, but research continues to suggest that in the main this is because of the following:

- HR specialist lacks the necessary skills and knowledge to analyse and interpret data and information (Lawler and Mohrman 2003; Williams *et al.*, 2008);
- there are concerns regarding the integrity, reliability and consistency of data (CIPD 2004; CIPD 2005).

There is a strong correlation between an organisation with good project management skills and knowledge and high satisfaction with a HRIS implementation (CIPD 2004; CIPD 2005). According to Tansley *et al.* (2001), much of the success depends on nine key factors:

1   senior management support and commitment;
2   involvement of representatives from all potential user groups in the project team (e.g. senior managers, HR, IT, line management, employees);
3   provision of comprehensive range of information on both on the potential system(s) and their potential benefits;
4   suitability of the potential system to the industry/sector of work and the specific challenges its environment places;
5   suitability of the system to the organisational culture(s) and management style;
6   differences of operating systems/approaches within different organisational divisions – need for integration/business process review and redesign;
7   benefits vs costs;
8   potential uses of the system (automate/inform), and;
9   relationship between HR and HRIS strategy and policies.

Although HRISs are widely used to support the employee resourcing function within other sectors, the ways and levels of practice within construction organisations vary greatly (Raidén *et al.*, 2001, 2008). Empirical evidence of the use of HRISs in the civil engineering and construction sectors is also limited. Besides some survey work (Raidén *et al.*, 2001, 2008), the few studies undertaken tend to be located in non-UK organisations (see Ng *et al.*, 2001; Florkowski and Olivas-Lujan, 2006). Notably however, the limited research is united in that given the high mobile and transient nature of the construction workforce, HRIS offer a more reliable, accurate and accessible means of human resource planning, reducing labour turnover and targeted training and development (Ng *et al.*, 2001; Raidén *et al.*, 2001, 2008). Ng *et al.* (2001) found that this was done through seven major functions: (1) project management and control, (2) strategic planning, review and analysis, (3) employee profile, (4) employee performance, (5) human resource development, (6) payroll and accounting, and (7) information systems outside the company. These sum up to the key advantages of using a HRIS to support strategic HRM decision-making in construction identified by Loosemore *et al.* (2003: 109):

- easier provision of information to line managers, thereby enabling rapid resourcing decisions during projects;
- easier processing and control of employee records and performance data linked to reward systems (i.e. removing the need for managers to maintain unwieldy paper-based systems);
- a reduction in the workload of the personnel function, thereby lowering the head office overhead associated with the strategic HRM function.

Chapter 4 provides further discussion on the use of HRISs in construction organisations as a route to addressing some of the issues emerging from the research.

Change management

Many of the challenges within employee resourcing are concerned with change. Thus, the fourth strategic HRM objective of employee resourcing activities seek to fulfil is change management. Change management as an academic subject is an area of significant research and many organisations have dedicated departments or teams managing their change programmes. The employee resourcing remit of change management is not to interfere with this but rather support an integrated implementation and application of change. The aim is to ensure that the significance of change in organisations is recognised and that changes are managed effectively (Taylor, 2005). This involves the management of processes through which the organisational structure and culture progressively evolve, such as recruitment and selection, team deployment and performance and career management. Thus, in the context of this book, change management refers to the means to achieve enhanced organisational effectiveness and individual development via a strategic and integrated employee resourcing activities. Employee involvement is one strategy organisations employ to encourage flexibility and change management.

Employee involvement

Employee involvement is aimed at achieving staff commitment and participation through increasing employee voice and decision-making power. In
essence, it is about increasing organisational effectiveness through manager
and employee collaboration and through sharing power and control. Kochan
et al.'s (1986) early work suggested that employee voice be addressed in
two ways:

- by providing opportunities for employees or their representatives to
  be engaged in decisions affecting their jobs and terms and conditions;
- by actively resolving disputes of interest.

Employee involvement has been a focus of attention for many years in the
HR field. In the past, terms like participative decision-making have been
used, but more recently the scope of involvement has been extended through
the concept of 'empowerment', a shift towards a greater emphasis upon
trust and commitment at the workplace (see Dainty *et al.*, 2002).

Despite the espoused benefits of empowerment to performance, it
remains a poorly defined term, and the literature provides a wide range of
competing definitions (see Geroy *et al.*, 1998). Empowerment is often
attached to popular concepts such as *Total Quality Management* (TQM),
Business Process Reengineering (BPR), *Teamworking* and the *Learning
Organisation* (Wilkinson, 2001). These all require increased participation
of employees as necessary component for improving the efficiency and
effectiveness of organisations (see Cook, 1994). However, they also demand
different approaches towards ensuring this. Empowerment has also been
widely criticised. For example, Wilkinson (2001) comments that it means
different things in different organisations, it is seen as an entirely new concept with no historical context, little discussion is available on its effective
implementation, and the context within which empowerment takes place is
largely ignored. Others highlight barriers to its implementation such as a
lack of management support (Wysocki, 1990 cf. Swenson, 1997), perceptions that it is linked to downsizing (Adler, 1993) or that it can create
inappropriate competition between teams which can detract from its objectives (O'Conner, 1990 cf. Swenson, 1997). Clearly, if it is treated as a term to
encompass the delegation or handing off of responsibilities without due
consideration to the process of managing the transition, empowerment is
likely to lead to little more than abandonment (Ghosal and Bartlett, 1998).
Thus, the challenge for managers wishing to implement empowerment is to
ensure that the 'soft' HRM rhetoric of empowerment does not mask a
'hard' reality of *someone else taking risk and responsibility* (see Sisson,
1994).

Through the growing importance and extending scope of employee
involvement, this has progressed towards a variety of more informal

practices. However, while in practice employee involvement now takes a variety of forms; Marchington (1995) categorised them into five groups (Table 3.7).

Different types of employee involvement are often found to coexist, particularly within organisations where it is a central element of the overall management style. Perhaps the most common types are downward communication, consultation and representative participation, and financial participation. These have been found a particularly effective way to managing change (Mabey *et al.*, 1998), improving performance (Cruise O'Brien, 1995), ensuring customer satisfaction and encouraging innovation (Wickisier, 1997). In sum, employee involvement can help organisations to ensure

Table 3.7 Five types of employee involvement (Marchington, 1995; Corbridge and Pilbeam, 1998: 332–334)

Type	Objective	Techniques
Downward communication	Managers to provide information to employees in order to develop their understanding of organisational plans and objectives	Formal and informal communications: reports, newspapers, videos, presentations, team briefings
Upward problem-solving	Utilise the knowledge and opinions of employees to, e.g., increase the stock of ideas within the organisation, encourage co-operative relationships and legitimise change	Suggestion schemes, TQM and quality circles, attitude surveys
Task participation	Encourage employees to expand the range of tasks they undertake	Job rotation, job enrichment, teamworking, empowerment, semi-autonomous work groups
Consultation and representative participation	An indirect form of employee involvement, aiming to support effective decision-making, air grievances, 'sound out' employee views on organisational plans	Joint consultation, discussions between managers and employees/ their representatives
Financial participation	Relate the employees' overall pay to the success of the organisation with the assumption that employees will work harder if they receive a personal financial reward from the organisation's success	Profit-sharing schemes, employee share ownership plans

employee commitment through balanced psychological contracts, flexibility and career management by incorporating the individual employees' varying needs and preferences into their planning and policy making processes. Sophisticated technology applications offer significant potential to support this.

In construction organisations, evidence of employee involvement/ empowerment is currently poor (Druker, 2007). This is despite the governmental suggestions and strong positive message communicated by research in the area. Advocates of employee involvement have tried hard to press for the concept to be embodied as an integral aspect of contemporary construction management. For example, Tener (1993) argued that 'the profitability and competitiveness of a firm are driven by, *more than any other element*, the ability of the organisation's leadership to empower its people to perform to their maximum potential' (emphasis in original). Also, Long (1997) identified the concept as a positive driver towards enhanced employee performance and corporate success within the industry. Indeed, the widespread recognition of the negative implications of fragmentation on the performance of the sector has led to attention being paid to developing new working practices and structures which encourage the integration of project activities and participants in the design and construction process. Thus, more recently, Dainty *et al.* (2002) suggested that if used selectively, it could play an important part in helping construction organisations to address increasing performance demands whilst mitigating the negative effects of the fragmented project delivery process'. However, they also recognised the potential barriers to successful implementation of employee involvement within construction organisations and proposed the concept to be best implemented where flatter management/organisational structures, formal support networks and devolved lines of responsibility with regard to project production delivery exist (Dainty *et al.*, 2002; Loosemore *et al.*, 2003).

Psychological contract

Another central theme in change management in the context of the employment relationship is the concept of psychological contract. Rousseau (1994) defines the psychological contract of employment as the understanding people have, whether written or unwritten, regarding the commitments made between themselves and their organisation. This infers a mutual expectation of commitment from employer and employee, or a two-way exchange of perceived promises and obligations (Guest and Conway, 2000). The psychological contract can be positioned anywhere along a continuum bounded by two distinct theoretical types. These are as follows:

- *Relational contracts* – long-term, open-ended relationships within unitary organisations which lead to the exchange of loyalty, trust and support.

- *Transactional contracts* – short-term relationships set within pluralistic organisational contexts and characterised by mutual self-interest (Rousseau, 1995).

Despite its central role in understanding the employment relationship, construction community has paid little attention the psychological contract construct. Dainty *et al.* (2004c) examined the nature of construction project managers within the following framework. They suggest that regardless of where the psychological contracts sit within the transactional-relational continuum, it should be seen as interactive and dynamic. The employer and employee continually inform, negotiate, monitor and re-negotiate (or exit) the employment relationship. From a functional perspective, psychological contracts accomplish two tasks: they help to predict the kinds of outputs which employers will get from employees, and what kind of rewards the employee will get from investing time and effort in the organisation (Hiltrop, 1996). Qualities central to the psychological contract include individual differences, interpersonal interaction, motivation, leadership and management style, group/team dynamics, change and empowerment. A breach, break or violation of the psychological contract will have negative impacts on its qualities. These may include reduced trust, job satisfaction and commitment to remaining with the organisation and the withdrawal of some types of employee obligation (Robinson and Rousseau, 1994; Hiltrop, 1996; Lester and Kickul, 2001). Consequently, understanding the content of psychological contract is the key to understanding the factors which lie behind employee turnover. As a result, their work identified that the traditional relations are now being strained by widespread organisational expansion and flattening organisational structures. In response, they suggest that construction firms must develop HRM policies which emphasises career development and recognises the contribution of the individual, rather than rely on pluralistic solutions. Chapter 4 discusses this further.

3.4 Models of employee resourcing in construction

So far, the employee resourcing function has been considered from a general HRM perspective after Taylor (2002, 2005). He identified four central elements to the function: staffing, performance management, HR administration and change. The specific HRM activities were located within these elements (e.g. human resource planning, recruitment and selection, team deployment and exits under staffing) to provide a comprehensive, holistic framework for discussion. To provide an explicit focus for discussion on the issues specific to construction organisations, this section considers four employee resourcing models developed for/within the industry. These are Serpell and Maturana's (1995) SAGPER, Trejo *et al.*'s (2002) framework for competency and capability assessment for resource allocation, Loosemore *et al.*'s (2003: 84) model of the competing pressures and interdependent functions involved in the resourcing of project teams and a model of the employee resourcing cycle in the industry (Raidén *et al.*, 2002a, 2002b).

SAGPER

The SAGPER model[1] presents a knowledge-based system methodology for supporting management decisions and improving human resource management in construction projects. This is achieved by providing managers with suitable information about employee motivation and work satisfaction on a construction site. Serpell and Maturana (1995) developed the model in response to increasing competition, more technically complex projects, client demands and a shortage of qualified labour in the Chilean construction industry. Their research showed that one of the main limitations faced by construction managers for managing human resources was the almost absolute lack of information about most of the aspects related to personnel. The model was to improve this via effective collection and analysis of required data with the suitable steps of the process being automated. On the basis of this evaluation, recommendations for improvement were to be generated and presented to the managers, whom were assumed to hold the responsibility for their implementation. From the outcomes of this data collection and analysis, recommendations and implementation cycle, the learned experience was to be fed back to the data collection and analysis process as well as the company personnel specialists. The personnel specialists were assumed to hold responsibility for the design and quality of the data collection instruments. Figure 3.4 shows the general architecture for the SAGPER.

SAGPER suggests significant improvements to construction human resource management via structured decision-support with the potential benefits of acquiring empirical evidence on site-based HRM issues, storing labour management expertise and making crucial knowledge available to a large number of managers. The model utilises expert system technology to support its automated components. This feature contains a set of rules, which evaluate the level of satisfaction (and dissatisfaction) among the specified site operatives on the basis of the input data. This satisfaction measure is calculated from a combination of 11 factors:

- materials availability
- responsibility
- work stability and pressure
- relationships with supervisor and managers
- communication
- compensation
- safety conditions
- work facilities
- recreation
- food.

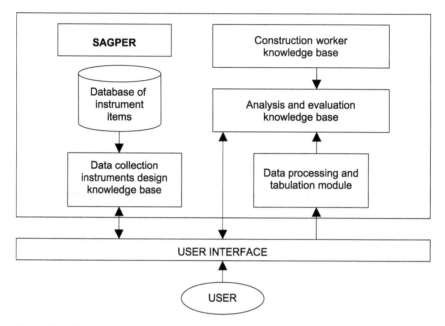

Figure 3.4 General architecture of SAGPER (Serpell and Maturana, 1995).

As such, the practical applicability of the model to a wide range of situations could be constrained by the limited number and range of factors included in the analytical process. Furthermore, the model being entirely project/site focused, its applicability to managing the varied employee resourcing activities organisation-wide is negligible. Finally, the use of the model is limited by its focus on site operatives, who are commonly employed by sub-contractors on projects managed by large contractors in the UK, and evaluating the satisfaction/dissatisfaction rates of these site operatives, rather than managing the complex and diverse employee resourcing activities in an integrated manner as suggested most appropriate by the strategic HRM approach.

Framework for competency and capability assessment for resource allocation

Trejo *et al.*'s (2002) framework for competency and capability assessment for resource allocation is a process flowchart that focuses on matching individual employee competencies/capabilities to the requirements of the organisation. This model was developed under the assumption that a systematic approach to determining accurate predictions of the organisational strengths and weaknesses will enable managers to establish optimum resource allocations. This approach allows no concern for what the individual employee could offer to the organisation beyond the immediate requirements, for example, to help improve/widen organisation's opportunities. The design of the framework was based on a review of other models which have the sole focus on technical capabilities, and therefore exclude factors such as interpersonal skills or client preferences. Thus, the framework lacks concern for 'soft' people issues. Its purely 'hard' skills and abilities application does not take into account personal aspirations, psychological contracts or other important themes within people management discussed before.

Model of the competing pressures and interdependent functions involved in the resourcing of project teams

The model of the competing pressures and interdependent functions involved in the resourcing of project teams is shown in Figure 3.5.

> This model illustrates the complexity of the construction project-resourcing environment . . . [and] reveals the many potentially competing objectives which the resourcing process must meet.
>
> (Loosemore et al.'s (2003: 84)

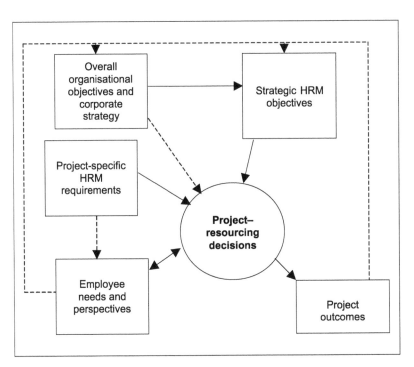

Figure 3.5 A model of the competing pressures and interdependent functions involved in the resourcing of project teams (Loosemore *et al.*, 2003: 84).

As such, the model usefully fulfils its purpose. However, regardless of the concern for project-resourcing *environment*, it fails to take into account the crucial impact of the environment external to the organisation, which, as discussed in Chapter 2, has a fundamental effect on way the employee resourcing activities are managed. In addition, Loosemore *et al.*'s model fails to recognise the impact project-resourcing decisions have on the project specific HRM requirements, overall organisational objectives and corporate strategy and strategic HRM objectives. Links illustrated in the model show only one way relationships between these aspects. This closed-systems view appears unable to contend with the multiplicity of environmental influences which impact on resourcing decision making.

Understanding the employee resourcing cycle in construction

In an attempt to model the employee resourcing cycle, Figure 3.6 highlights the key employee resourcing activities and the internal and external factors that influence the function within the industry. The employee resourcing specific strategic HRM objectives of staffing, performance, HR administration and change management form the central core of the cycle. Learning and development and employment relations are closely linked to the individual activities within the function. The ultimate goal of the function is the continuous achievement of organisational goals at a minimum risk while maintaining balanced psychological contracts.

Factors internal and external to the organisational context influence the effective management of the function. The internal influences within the model (Figure 3.6) include *the organisation's strategic choice in terms of strategic HRM*, which often affects higher level issues of employee resourcing, development and relations. In construction this may mean an organisation opting to employ staff only on fixed-term temporary contracts, invest heavily in training or devolving HR responsibility from HR practitioners to line management. *Organisational structure* is another significant internal influence. This may mean, for example, adopting a hierarchical structure or differing functional and hierarchical reporting lines, so called matrix organisation. *Organisational culture* can influence the aforementioned factors as well as the strategic HRM practices. This is more difficult to manage and control, as it is difficult to describe and measure, and varies between organisational departments and divisions. Finally, *factors central to the individual employees within the organisation* have an impact on employee resourcing. Similarly to the organisational culture, employee needs and preferences may not have such a direct or tangible effect on the strategic HRM practices as the organisational strategic choice or structure, although they should influence the developmental stages of strategic HRM processes. The processes should serve the needs of the managers in managing the employees and the needs of the employees' in involving them in the process.

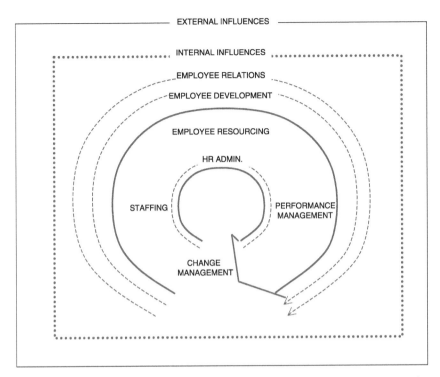

Figure 3.6 Understanding the employee resourcing cycle in construction (after Raidén *et al.*, 2002a, 2002b).

Ultimately, however, the internal factors contribute to the strategic business objectives of employee resourcing: the continuous achievement of organisational goals at a minimum risk while maintaining balanced psychological contracts. As was established earlier, understanding the psychological contract is vital to understanding employee responses to the employment relationship. Thus, to ensure maximum productivity, it is crucial that the individuals' preferences and expectations are taken into account throughout the resourcing process.

In addition to the factors internal to the organisation influencing the strategic HRM processes, several factors external to the organisation affect the way strategic HRM practices are organised. The fluid and dynamic environment of the construction industry presents a particularly problematic context for effective employee resourcing. The challenges include those that apply to construction industry specifically (see Chapter 1) and those, which apply to all business sectors. Common throughout all business sectors are technological, legislative and demographic changes; changes in peoples' values and beliefs, quality standards and expectations; and changes in the economic/labour markets.

Further challenges to the process introduce the need to balance these competing external, organisational, project and individual priorities and needs, both at strategic (long-term) and operational (short-term) levels. As discussed in the earlier parts of this chapter, the current employee resourcing practices often rely on the personal assessments of line management, which have the potential for inconsistencies, poor allocation decisions and hence, disillusioned employees through the violation of the psychological contract. Due to the project/organisational needs predominating the process, this has the potential to contribute to increase employee turnover, and hence, to contribute to the overall inefficiency of the industry. This places extreme demands on both HR departments and line managers, and requires a flexible approach to the employee resourcing function in construction organisations.

3.5 Summary

This chapter has discussed the three distinct but interrelated functions of strategic HRM: employment relations, which provide the overarching management style and philosophy for managing people in organisations; learning and development, which facilitate organisational and individual improvement via training and development initiatives; and employee resourcing, which focuses on staffing organisations, managing performance, HR administration and change management. Within these elements contemporary themes such as the learning organisation, flexibility, work–life balance, careers, HRISs, employee involvement and the psychological contract were discussed with attention on the specific challenges of the construction industry. Finally, four models of employee resourcing were reviewed. The next chapter will look at in-depth empirical research on employee resourcing practices among large contracting organisations in the UK and offer critical discussion on the findings.

Note

1   SAGPER by its Spanish acronym.

4   A review of current practice

Over the past decade, strategic HRM has developed as an effective approach to people management (see Chapters 2 and 3). It is now a widely adopted structural and operational basis for organising the recruitment and selection, staff retention and performance improvement functions of many public and private sector organisations. In the construction industry however, attention has been much less pervasive. There are important characteristics of the industry that challenge the adoption of such concepts. Chapter 1 discussed the dynamic and complex project-based nature of work in the industry: most construction projects are unique organisations designed and built to meet particular client specifications. They are often awarded at short notice and rely on transient workforce due to site location. Clients are demanding and the industry workforce is heavily male dominated.

The site-based operations require construction organisations to set up temporary organisational structures at dispersed geographical locations, frequently at a distance from central management. Thus, the project team forms the focus of working life, operating with a significant and necessary degree of independence. The changing requirements of each project also necessitate the formation of bespoke teams each time a new project is awarded. Since the time available between contract award and the mobilisation of the project is usually extremely limited, planning is difficult and the workforce must be highly mobile and flexible. The short-term timescales force quick decisions on fundamental aspects of organisations' working life and operations. As the literature suggests, for this reason many construction managers focus on the achievement of financial, programme and quality outcomes over broader project performance criteria.

The male-dominated, 'macho' culture of the industry and typically long-working hours have a negative impact on construction organisations' ability to recruit and retain talented workforce, which are essential for satisfying the steadily increasing expectations of the clients.

This chapter examines the HRM practices of seven large construction organisations in the UK. It presents a critical review of the employment relations and learning and development issues before focusing on the specifics of employee resourcing that are central to the success of strategic HRM in construction. That is, allocating staff to projects by simultaneously meeting the (often competing) needs and requirements of the organisation, the project and individual employees. Project case studies are used to illuminate the contextual and contingent nature of resourcing decisions. Where promising practice was observed, it is highlighted to provide suggestions for addressing the challenges inherent in the resourcing process. The aim here is not to provide guidance in a normative or prescriptive manner, but to stimulate thinking around how such challenges can be addressed. Such ideas would, of course, have to be appropriated and enacted in a contingent way if other companies are to benefit from them.

4.1 Synopsis

One central theme runs through the discussion of the research findings: many of the case study organisations operate in a style that is broadly representative of the IR/personnel type approach to people management. This is in line with Druker *et al.*'s (1996) conclusions which suggest that the industry adopts personnel management style to the effect that the industry has retained a short-term approach to the management of people. However, aspects of strategic HRM are apparent in the general management style, such as the aim to go 'beyond contract', need for a 'can do' outlook and drive on business needs, flexibility and commitment as shown in Section 4.3. The organisational strategy and values of the primary case in particular support this. Especially the company's commitment to empowerment and training and development, together with the friendly and family-oriented culture arguably provide a firm footing for a more strategic approach towards HRM.

Many of the case organisations are also well positioned in relation to line management capabilities. Central to the concept of strategic HRM are business–customer relations and devolvement of key management activities to the operational managers, which are both crucial to the way the construction organisations are managed. The devolution of key management activities to operational line managers was a particular feature, a trend also noted by Druker and White (1995, 1996) as a characteristic of more strategic HRM within the industry as a whole.

The greatest employee resourcing problem seemed to stem from the diversity of local practices found at a divisional/project level. The organisation level strategic intention was generally very positive. Managers attempted to plan for the human resource requirements via human resource planning (HRP) activities. An informal culture was particularly apparent within the primary case. This emphasised divisional/departmental loyalty and close working relationships between staff and their line managers. Although useful in enhancing local commitment, it failed to translate the strategic intentions of the organisational level into effective managerial practice in projects. Employees moving from one project to another, or temporarily between divisions, were met with radically different sub-cultures and ways of working.

4.2 Strategic choices and organisational priorities

The focus of the primary case study organisation's corporate strategy was to deliver leading edge complete construction services through partnership approach. This was developed into an organisational vision and values which focused on building long-term relationships with customers, employees and other stakeholders. The vision and values sought business excellence through innovation, continuous improvement, employee empowerment and a professional, open and honest approach to conducting business. Fairness and trust were at the heart of their operations strategy.

Furthermore, the people statement in the Annual Report and Accounts read to the effect that the company recognised employee contribution to the organisational success. The HRM strategy was designed to follow best practice and encourage an active learning environment. Structured training and development interventions were provided to all employees to motivate them and maximise their job satisfaction and contribution to the business.

The company proudly advertised itself as an equal opportunities employer, with broad policy and procedures in place across the group in recruitment, benefits, training and advancement. Cultural differences and diversity were respected. Applications from ethnic minorities and the disabled were encouraged in job advertisements. Furthermore, the people statement said that their employees' work–life balance and privacy were very important to the organisation.

A competency-based annual appraisal system that included a personal development plans for all employees was in place. The organisational aim was to have a sufficient mix of people with the right competencies to meet the current and future needs of the business plan.

The company achieved the Investors in People standard accreditation in 2001. This provided the focus for the corporate training and development strategy. Continuous progress ensured new training schemes were embedded into the business and the spirit of lifelong learning maintained. In the long term, the company aimed towards 'fully qualified' workforce by linking existing and new training programmes to national vocational qualifications (NVQs).

As the organisation valued open, honest and constructive communications throughout the group and its wider supply chain, employee communication continued to be improved. Staff at all levels were consulted on matters about the progress of the company. Also, other issues that may be of interest or concern to employees were discussed openly. Quarterly employee briefings, half-yearly updates on financial results of the operating companies of the group and an annual employee survey together with twice-yearly company magazine and an employee newsletter were some of the formal mechanisms for employee communication. Informally the company operated an open door policy.

This clearly suggests a significant degree of senior management commitment to proactive and strategic people management practice and continuous learning and development, equal opportunities, work–life balance and employee communication.

Delivering the strategy, vision and values form the focus of the human resource plan, which is formulated annually as part of the overall business plan. The plan is distributed to the divisional directors and senior managers whom hold the responsibility for its implementation, with the support of a central HR department.

Devolved HRM and the organisational structure

The company's strategic choice in terms of operational HRM was to devolve many of the responsibilities to line management. Although the company had a main board level HR director, HR staff generally had little influence at divisional/project level. The HR director worked as a strategic link between the regional profit centres. He had a main board position, with the responsibility of looking after the company's HR and training budgets. These funds were allocated between the regions as necessary.

The company had four designated staff for dealing with strategic HRM-related matters. This included:

- the Managing Director's (MD) Personal Assistant (PA), who looked after the executive recruitment and selection process;
- quality, Environment and Training manager, who administered and monitored the in-house training database;
- a Payroll Administrator, who looked after the operative recruitment and selection administration and payroll;
- a Graduate Recruitment and Development Officer, who managed the graduate recruitment and development programmes.

The company's organisational structure was broadly hierarchical despite the dynamic nature of the industry. This was apart from a limited number of small pockets of matrix management, which were found within the quantity surveying staff (under 'procurement' in the following organisation chart – Figure 4.1).

Since the day-to-day responsibilities for HRM were devolved to operational line managers the role of the HR specialists was found to be purely advisory. It was the responsibility of divisional directors and senior managers to ensure that the operations run smoothly, targets were met and the personnel involved in projects were looked after appropriately.

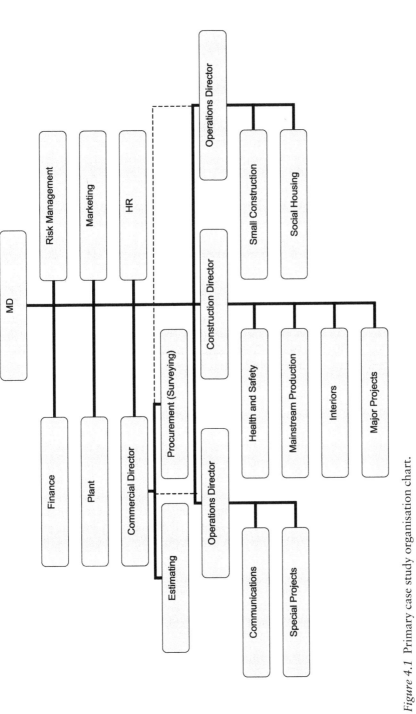

Figure 4.1 Primary case study organisation chart.

Many managers viewed the HR function as intrusive and unnecessarily bureaucratic:

> [They are] always chasing for information and records, where I have a good system here locally.
>
> (Divisional MD)

Operations managers did not acknowledge or understand the organisation-wide benefits of having a centralised strategic HRM support or information database. This may reflect the operational autonomy with which the divisions operated, although this undermined the potential to strategically redeploy staff across the business.

The employees had minimal contact with HR staff. Line management formed the main source of HR information for project-based staff. The only HR-related contacts the employee respondents mentioned were meetings with the company Health and Safety Officers. These were employed under the personnel function, however, their role focused solely on site safety inspections and related guidance.

In the secondary cases too, the HR specialists' role, although strategic to the extent of providing policy and procedural support, tended to revolve around reactively 'fire-fighting' problems as they occurred and dealing with their consequences, rather than proactively preventing the problems from occurring in the first place. For example, problematic equal opportunities issues with the potential of industrial tribunal action were referred to the department rather than managed appropriately within initial recruitment and selection and performance management procedures by the operating divisions. One HR manager (from company A) explained:

> Ironic this is really, but HR is dealing with the wrong end of the business. Our main problems are associated with terminations: industrial tribunals, compensation and lots of hassle. HR does not get an opportunity to input to line management decision-making as to the suitability of the people they are taking in. If you are trying to run a successful business, it is the people that you bring in that are critical. Then we wouldn't need to get involved in so many negative exit situations.

In some of the other organisations (companies C and D), the role of HR was seen as that of a 'negotiator' and a 'coach'. As in the earlier example, HR was called to interfere and work out solutions to problem situations. Managers also turned to HR for support and advice on dealing with many sensitive issues in the workplace, such as work–life balance conflict. Devolvement of operational HRM responsibilities was common and so line managers formed the main point of contact for project-based staff in terms of promotional opportunities, project/team deployment issues and other

HR queries. Employees were said to approach HR most commonly when they required information on leave, grievance procedures or developmental opportunities.

Organisational culture

In line with the company's 'People Statement', the vast majority of the primary case respondents described the organisational culture as friendly, open and family orientated with two-way communications at the heart of the operations. However, some employees referred to the existence of an internal 'old boy's network'. They suggested that newcomers were 'tested' by longer serving members of staff at site level. Only when the new employees had proven themselves to their colleagues were they accepted as members of the team.

One of the secondary case respondents referred to the existence of a 'drip feed' culture with regard to implementation of good people management practice:

> We are backwards as regards to HR and it is frustrating not to be able to implement HR initiatives that you read about in journals and the press. E should be adopting similar principles ... but you have to drip feed everything in: you put an initiative forward to the board, get it rejected, change it slightly and put it forward again in six months time ... over time change happens, but unfortunately most likely as a result of regulatory/legislative change.
>
> (HR Officer, Company E)

Employment relations policy and procedures

At policy level, the primary case organisation had established a firm standing on employment relations. Equal opportunities, grievance, dismissal and other related policies were well in place, as indicated by the company's People Statement. However, many of the policies/procedures had rarely translated into effective practice at project/site level. Many managers and employees were unaware of the existence of such documentation, or even its possible location should a need for reference arise. Despite this, employment relations were not found to be a significant issue. The friendly organisational culture and individualistic management style gave many a feeling of confidence in fair practice. Trades union density at professional level was minimal. This is not unusual given the generally low level of representation in the industry, even among the operative workforce (Druker, 2007).

Pay was negotiated individually within wide scales and a standard benefits package was applied to all staff. Annual bonus was assigned on the basis of the company's financial performance and each employee's individual achievements. All respondents were happy with the bonus arrangements. However, few complained that they were only remunerated for their contracted working hours despite committing many more in order to secure successful project outcomes. Overall, the company approach to employment relations indicate promising platform for developing strategic HRM excellence through effective employee-resourcing strategies.

Management style

The research interviews with the senior managers of the primary case study organisation included an assessment of the respondents' management or leadership style. Overall, only a few managers showed balanced, strategic HRM type leadership. However, two respondents indicated strong tendency to focus on people. This was a surprise taking the demanding operational requirements of construction industry into account.

Some important details were revealed. Firstly, an examination of the data across the group (the senior management team), rather than separately by individual managers, suggests a relatively balanced overall decision-making. While some managers adopted balanced, strategic approach and others had distinctively people-centred approach, some were clearly concerned with production-oriented approach. This combination makes the senior management team together a strong and balanced group, although individually the respondents are likely to prefer highly different solutions.

Secondly, responses related to organisational culture and teamwork were very positive along the balanced, strategic HRM approach. Only one respondent held strongly 'production-oriented' views. This may suggest that the organisation having an open, no-blame culture where innovation and new ideas are encouraged. The responses on conflict support this. The management team's approach to conflict handling is unmistakably people centred. However, in contrast to these reflections on organisations culture, the assessment revealed a rather production-oriented response on showing emotion at work.

The distinct production orientation towards emotions at work combined with open, no-blame culture in the organisation suggests an environment of 'hard people orientation'. Focus and effort is placed on people when performance is required, but much colder, production-oriented approach is adopted when the 'softer' emotional aspects of work relationships are in question. Overall, these results confirm a promising platform for developing strategic HRM excellence through effective employee-resourcing strategies.

The secondary case data suggested that an operations-oriented management style is as a characteristic of the way the industry operates. In company B, the HR manager described the attitude some of their line managers adapt to personnel selection and retention as 'well, if it doesn't work out, just let them go'. A short-term outlook was also apparent in planning. For example, HRP was often focused on seeking solutions within one year forecasts, which in terms of strategic planning is relatively short-term (HRP is discussed more extensively in Section 4.5).

Organisational strategic priorities and operational project requirements

The managerial respondents highlighted a number of factors (N=92) that were considered important to be taken into account in the resourcing decision-making process (Table 4.1). These included HRM/organisational issues, which affect the organisation as a whole and are managed at strategy/policy level within the higher senior management positions. Project/team-centred variables were also identified. These consist of specific activities and factors that have direct impact on the project process and outcomes as well as desired employee qualities including factors the managerial informants stated as necessary, or advantageous, for carrying out the tasks involved in most construction roles.

Closer examination of the factors suggested that a thematic grouping facilitates clearer understanding of the variables due to the wide variety of factors involved and the interlinking nature of the HRM/organisational and project/team issues. Unsurprisingly, many of the factors managers highlighted as important to the resourcing decision-making process focus on operational effectiveness and client relations. Softer cultural aspects form an equally important component of the overall picture. Five major themes are significant: team/project environment, employee involvement/communications, careers, learning and development, and organisational planning.

Team/project environment

In relation to the team/project category, several factors were associated with notions of team spirit and relationships, which highlight the importance of collaborative teamwork/partnership culture. For example, several project managers emphasised the importance of all communication to be non-controversial. The respondents referred to 'company culture' in that all correspondence must be constructive and positively worded. Senior managers encouraged early intervention and preferred any situations to be solved at middle management tiers, close to the problem. This was believed to be an effective way of tackling problems. Any disruptive influences were removed rapidly, with little tolerance for attitude problems. An overall spirit of collaboration and working together was said to help in managing the team–client relationships for mutual benefit. More procedural variables, such as induction, problem-solving and clarity of roles and responsibilities, supported this view. In addition, the need for a manager to know his/her staff, the best use of employee skills, talents and abilities, and balanced team factors emphasise the importance of effective team formation and deployment. Although these were important for decision-makers as well as the employees, business demands often made it difficult to incorporate them into the decision-making process explicitly. Clearly the managers were also concerned with the commercial aspects (such as surveying, profitability, quality), client satisfaction (through giving value for money, client involvement and ensuring client preferences are taken into account at every stage of the project) and sub-contractor performance (including health and safety). Strong leadership was identified as being essential for achieving this kind of working environment.

Employee involvement/communications

The employee involvement/communications group shows the significance of appropriate management style to resourcing decision-making. An open and approachable attitude is emphasised through effective two-way communications, delegation and the requirement for recognition of individual employee contribution. An 'open-door' communications policy is suggested as the appropriate procedural solution together with short and direct links through management structure.

Careers, learning and development

The careers and learning and development themes draw attention to organisational development and continuous improvement. A clear link with knowledge management is apparent in cross-project learning, job rotation and graduate development. Management style is also further emphasised. Transparent progression opportunities and succession planning focus on staff retention and achievement of organisational goals in the long-term via career development, fast track progression and taking on trainees. These provide a foundation for a 'learning organisation' culture (see Section 4.4).

Organisational planning

Much of the organisational context focuses on operational effectiveness: gaining repeat business, quality assurance, working within legislation and industry regulations. Some strategic considerations are included in terms of broadening the business, stabilising/increasing financial turnover, integration within larger groups, marketing and business development. Workload and softer, cultural factors direct attention to effective HRP. These emphasise the importance of long-term planning, organisational flexibility and management of change via organisational culture founded on trust, openness, partnering, empowerment (employee involvement) and individualistic management style. Specific issues of concern to the managers are shifting emphasis from civil engineering towards building, balancing the ratio of freelance and permanent staff, keeping employee turnover manageable and encouraging continuous improvement.

Other factors consisted of issues to do with remuneration (equal terms and conditions for staff, pay), work–life balance (travel, job satisfaction/enjoyment) and desired employee characteristics. Remuneration and work–life balance highlight factors contextual to the employee resourcing process. The desired employee characteristics outline variables directly relevant to staff recruitment and selection. These included several motivational qualities, such as determination to succeed, self-drive, imagination, 'can do' attitude, adaptability, flexibility, professional attitude, task orientation, people skills and ability to make quick decisions, together with a few pointers to the need for appropriate qualifications, technical competence and multiskilling.

Table 4.1 Organisational/project factors and desired employee qualities

HRM/organisational factors	Project/team factors	Desired employee qualities
• Transparent progression opportunities for employees – good retention tool • Succession planning • Career development (achievable aspirations) • Fast track progression • Taking on trainees • Organisation (and employee) development via training plan – improve weaknesses • Graduate development • Workload • Continuous improvement • Flexibility • Organisation culture and spirit • Maintaining the culture • Trust • No-blame culture • Openness • Communal breaks • Empowerment, people feel they have a contribution • Partnering • Broaden business in strategic terms • Two-way communication • Work within legislation/regulations (e.g. TUPE) • Shifting emphasis from civil engineering focus toward building (organisation level) • Repeat business • Individualistic management style • Longer-term planning • Balance between agency and permanent staff ratio	• Make best use of employee skills, talents and abilities • Project performance • Profitability • Efficiency of site • Health and safety • Quality • Time • Balanced team, blend of individuals, mix of personalities • Good/strong leadership: right leader (e.g. project manager) on team • Team spirit • Team relationships • Team–client relationships (highly important) • Sub-contractor relationships • No controversial correspondence • Client satisfaction • No attitude problem • Manager to know his/her staff, their abilities and needs (e.g. training needs) • Sub-contractor performance • Client preferences • Client involvement • Commercial aspects (surveying) • Effective communication, both internally and externally	• Highly motivated • Determination to succeed • Self-drive • 'Can do' attitude • Quick decisions • Multiskilling and ability to tie in various professions • Adaptable, ready to react on deviation required • Imagination • Flexible, broad people • Technical capability and competence • Professionally qualified • Professional attitude • Mature, trades background • Willing to travel • Task orientation • People skills

continued

Table 4.1 (continued)

HRM/organisational factors	Project/team factors	Desired employee qualities
• Short and direct links through management structure • Managers, approachable and accessible • Delegation • Employees to get recognition from management and colleagues • Open door policy • Manageable employee turnover • Job rotation • Quality assurance • Stabilise/increase turnover (£) • Integration within larger group (following the merger) • Marketing and developing business • Equal terms and conditions for staff • Work–life balance (good employer scenario) – facilitates retention • Induction: bring people in the company way of working, integrate quick and give opportunity to perform • Employee pay – when right strong retention factor • Managing change • Job satisfaction and enjoyment	• People not chasing the same ball – clear roles and responsibilities • Minimum travel – can affect performance/ employee health, facilitates retention • Disruptive influences removed fast • Singularity and focus (re: teams) • Situations be solved at middle management tiers • Tackle problems effectively • Give value for money • Cross-project learning • Effective information flows	

In summary, 92 distinct but interrelated HRM/organisational and project/team-focused factors needing to be taken into account in the employee resourcing decision-making process were identified within five key themes: team/project environment, employee involvement/communications, careers, learning and development, and organisational planning. These included variables that affect the organisation as a whole, those that have direct impact on the specific project processes and outcomes, and qualities necessary or advantageous for carrying out the tasks involved in most construction roles. Many of the factors identified are context specific and so are expected to change over time. For example, one of the current HRM/organisational priorities was to establish equal terms and conditions of employment for all staff. This is a direct result from the organisation's recent merger. Once the new staff have been integrated within the organisation, such a contractual issue is likely to be of lesser concern at strategic HRM level. Ideally, facilitating the achievement of balance between the HRM/organisational, project/team and individual employee factors in the resourcing decision-making forms the central focus. Little evidence of the managers considering the employee needs and preferences were found within the sample. The next section explores the individual needs and preferences of the employee respondents.

4.3 Employee needs and preferences

One of the objectives of the study was to compare and contrast the organisational strategic priorities, operational project requirements and individual employee needs and preferences. The organisational strategic choices, priorities and operational focus are explained earlier. This section examines the employee perspective in terms of their needs and preferences in relation to organisational resourcing decision-making. A discussion on the rank order of nine key variables (by different respondent groups and individuals) is followed by an analysis of the factors extracted from the employee interview data.

Good team relationships are most important

The analytic hierarchy method questionnaire provided an initial indicative rank order as to the importance the employee respondents placed on a range of employee resourcing related factors derived from literature. Since the questionnaire was administered within the interviews, a significantly high response rate of 88.6% was achieved (for further detail on the research methods see Section 1.2).

Overall, good team relationships (70%) together with personal and/or professional development (58%) and gaining broad and/or specialist experience (56%) were ranked as the most important factors. These were closely followed by work location close to home/maintaining work–life balance (52%) and promotional opportunities (51%). Project type (e.g. size, complexity, etc.; 47%) and experience in working under different procurement systems of contractual forms (43%) were positioned only just below the 50% mark. Interestingly, despite personal and/or professional development scoring second highest in importance, training opportunities were ranked as the second least important factor to be taken into account in resourcing decision-making by employees. It is likely that this is due to the company placing great importance on encouraging and operating a system of equal training opportunities for all staff and successfully implementing this (see Section 4.4). In addition, due to the culture of the organisation, which emphasises departmental loyalty and commitment to local practices (resulting in very little movement between divisions/departments), organisational division received a very low importance rating (11%).

Variations between the different respondent groups, in, for example, work location close to home/maintaining work–life balance, gaining broad and/or specialist experience and promotional opportunities were notable. Figures 4.2 and 4.3 illustrate the differences by comparison of job roles. Figure 4.4 highlights the differences in two individual accounts.

In examining Figure 4.4, it is clear that one of the respondents places very high importance on work location close to home/maintaining work–life balance (location), where for the other respondent this is totally irrelevant. In contrast, where one respondent places no importance at all to organisational division (division), the other respondent regards it relatively high priority. Indeed, with regard to all of the nine factors their opinions differ to a significant extent. This clearly illustrates how different each individual's needs and priorities may be in comparison to those of their colleagues. As discussed in Chapters 2 and 3, strategic HRM comprises a set of employment practices designed to maximise organisational integration, employee commitment, flexibility and quality of work (Guest, 1987: 503). The individualisation of the employment contract is one of the key developments in a move away from traditional welfare (IR/personnel management) oriented framework. Therefore, to align their practices with strategic HRM paradigm, it is crucial that organisations operate a system which is capable of taking into account the very specific individual needs and preferences.

In order to gain more in-depth understanding of the type of factors,

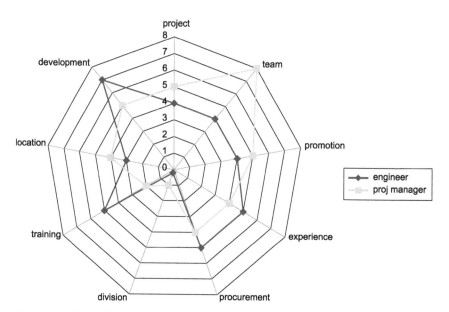

Figure 4.2 The differing resourcing priorities of engineers and project managers.

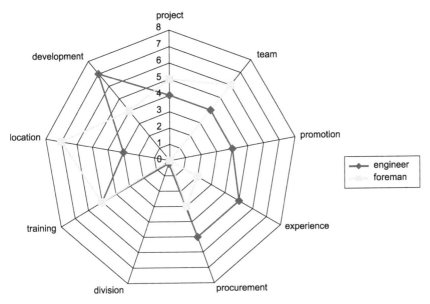

Figure 4.3 The differing resourcing priorities of engineers and foremen.

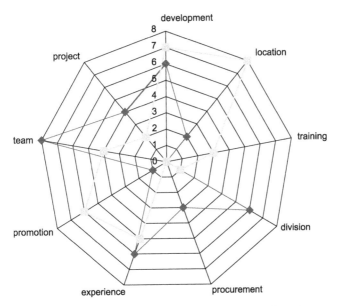

Figure 4.4 The differing resourcing priorities of two individual employees.

employees feel are important to be taken into account in the resourcing decision-making, and the rank order of these factors, the organisational context-specific employee needs and preferences were extracted from the interview data (N = 47) (Table 4.2). As with the organisational/project fac-

Table 4.2 Factors employee respondents highlighted as important to take into account in the resourcing decision-making

---

Progression
Promotion
Career (development)
Long hours (+working weekends)
Organisation culture
Training
Feedback on performance
Organisation structure
Employee Involvement
Communication
Feedback on progress
Recognition of contribution to organisation
Interdepartmental co-operation
Office location (and move of offices)
Travel
Work type
Appraisal
Team spirit
Project (deployment) opportunities
Moving between divisions/departments
Project (work) location
Pay
Employee suitability to project and specialisation in project allocation
Team redeployment (kept together)
Personal (and organisation) development
Staying (lodging) away
Work–life balance
Overtime
(New recruit/team member) induction
Health problems
Interest from top management/head office
Role and responsibilities
Communication on wider organisation issues
Discussion on expectations (both organisational and employees') – psychological contract
Team integration and co-ordination
Progress review
Team member selection
Nature of work
Job enrichment/enlargement
Horizontal moves
Best use of skills/talents
Recognition of individual qualities
Graduate development
Benefit preferences (e.g. car)
Procurement methods
Holiday
Comprehensive range of info on project

---

tors (Section 4.2) a thematic grouping of the factors is presented. Similarly, to the managerial responses, the employee responses show a relatively equal balance between procedural factors and the softer, cultural variables. In addition, the emphasis is on the same themes: team/project, employee involvement/communications, careers, and learning and development.

Team/project environment

The team/project factors focus on team member selection and the integration and co-ordination of the project activities. Team spirit was highlighted as the main cultural variable. Procedurally, employees considered a variety of project deployment opportunities, best use of their skills and talents, new recruit/team member induction and employee socialisation of crucial importance. The diversity of procurement methods were discussed to be diminishing because of client preferences but a preference for retaining some flexibility was indicated in order to secure wider ranging experience.

Employee involvement/communications

The employee involvement/communications-related variables centred on performance management systems, such as the appraisal and continuous feedback. The communications specific issues focus on the effective delivery of information throughout the organisation. Availability of comprehensive range of information on projects was of particular importance. Employee involvement and the recognition of individual contribution to the organisation as a whole integrated the cultural aspects into this theme.

Careers, learning and development

The careers and learning and development sections drew attention to progression and development within the organisation. Both upward mobility (promotions) and horizontal moves were desirable routes to personal development. Similarly, job enrichment/enlargement were discussed to encourage organisational learning and development. The nature of different roles and responsibilities, organisational structure and a variety of work were some of the organisational factors considered important for supporting holistic resourcing decision-making. Recognition of individual qualities and training and development programmes tailored to meet employee as well as organisational needs were other organisational support mechanisms together with a culture of co-operation, facilitating moves between divisions/department and explicit interest from senior management and the head office.

The central issue in relation to remuneration was benefits. Pay, overtime and holidays were also mentioned but the clear emphasis on benefits highlights the importance of recognising the different needs of individual employees and agreeing suitable arrangements. Work–life balance forms the focus in relation to the individual and industry characteristics. While project location was acknowledged as a characteristic of the industry beyond the control of the organisation (or employees) careful management of travel arrangements and especially lodging away as well as working hours were important to the respondents. Managers' sensitivity to the distance and frequency of the need to travel was appreciated.

In summary, the analytic hierarchy method exercise clearly established that employee needs vary between groups of employees categorised by job role. The questionnaire results also showed that the preferences and priorities vary between individual employees. The list of factors derived from the interview data added to this by specifying the type of needs and preferences employees feel important to be taken into account in the resourcing decision-making process. The factors were varied, ranging from organisational procurement methods to cultural issues such as team spirit and organisational co-operation. This included variables concerned with organisational role and division. Thus, the importance of organisational variables in addition to the factors concerned with the fulfilment of personal needs and wider career ambitions must be recognised in the deployment process. Furthermore, the nature of many variables involved suggests that the importance between as well as within the factors is likely to change over time as the individual employee's personal circumstances change. For example, typically a young, career ambitious trainee engineer with no family commitments would probably prefer promotional/developmental opportunities over work–life balance arrangements at the early stages of his/her career, where a more family-oriented and well-established senior engineer would be likely to place greater importance on organisational family-friendly leave policies and opportunities for flexible working.

The compatibility and conflicts between the employee perspectives, project requirements and organisational priorities

These variables extracted from the interview accounts of the managerial and employee respondents provide us an understanding of the wide variety of organisational/project and individual employee needs and preferences that are important to be taken into account in resourcing decision-making. A total of 139 variables were identified as important to be taken into account in the decision-making processes. Ninety-two of these were organisation/project related and 47 employee-centred. Five key themes emerged: team/project, employee involvement/communications, careers, learning and development and organisational planning.

A comparison of the factors the managers and employees felt important to take into account in the resourcing decision-making produced a list of 21 specific points of conflict, that is areas where the two sets of data were misaligned (Table 4.3). Five broad areas of compatibility emerged (Table 4.4).

Table 4.3 Principal points of conflict

Theme	Factor
Team/project	Team member selection
	Team integration and co-ordination
	New recruit/team member induction
	Best use of skills and talent
	Team spirit
Employee involvement and communications	Communication (overall)
	Comprehensive range of information on projects
	Feedback on performance/progress
Careers	Progression
	Promotion
	Horizontal moves
	Role and responsibilities
	Career development
Learning and development	Training
	Graduate development
	Personal and organisational learning and development
Other	Organisational culture
	Pay
	Personal health problems
	Work–life balance
	Travel

Many of the conflicts focused on deployment issues (team/project-related variables), employee involvement and communications, careers and learning and development. In addition, managers' and employees' perceptions differed on organisational culture, pay, travel, work–life balance and procedures for dealing with issue on personal health.

With regard to the team/project-related variables, the following activities need procedural improvement: team member selection, team integration and co-ordination, new recruit/team member induction and best use of skills and talent. Procedural change is also needed to improve organisational communication, particular with regard to employees having access to comprehensive range of information on projects and feedback on performance/progress. Career development must be enhanced via clearly defined roles and responsibilities and transparent progression, promotion and horizontal moves. Training and graduate development would also benefit from procedural change. Team spirit together with the broad remit of personal and organisational leaning and development incorporate much softer elements of resourcing into the decision-making. These are more challenging areas to address.

Table 4.4 outlines the areas of compatibility between the employees' and managerial views together with the positive features of each factor and their outcomes.

Overall, the analysis highlights the crucial importance of managing a few fundamental aspects of strategic HRM effectively. These are careers, remuneration, training and development, and employee involvement. The data also emphasises the importance of managing organisational culture and taking into account the personal circumstances of individual employees. As

Table 4.4 Areas of compatibility

Factor	Positive features	Outcomes
Formal training courses	Frequent Varied in nature Managers encourage attendance Regular updates	Well trained staff that have the required skills and qualifications to carry out their duties Employees realise organisational commitment and opportunities
Personal development plans	Training needs discussed Opportunity to highlight personal preferences Managers suggest/offer range of options	Personalised and tailored solutions Employee involvement and commitment Support for career management
Mentoring/coaching	More senior/longer serving members take (informal) responsibility to guide new recruits, team members and recently promoted personnel Part of the organisational culture at many levels	Informal support structure for personnel new to the organisation/role Enhanced internal relationships via employee participation in the organisational development
Work–life balance as a factor in team formation decision-making	Location one of the key criteria in decision-making Aim to 'rotate' staff travelling longer distances/staying away	Employees' travel requirements minimised Employee trust in managers looking after their staff Fairness of procedure
Individualistic management style	Management approachable and accessible Know their staff and their skills personally Open forum for discussion/grievances Genuine aim for good people management practice	Positive foundation for future opportunities via development of HRM practices Open communications Employee trust in managers Close relationships between managers and their staff

expected, the list includes factors specific to the construction industry: team formation and travel.

Although some of the HRM functions highlighted, such as remuneration, closely relate to the resourcing activities, there is no direct route to managing these aspects via the employee resourcing processes. Thus, these are considered contextual factors, which are essential for the process but cannot/are difficult to influence. They must be dealt with separately from the resourcing decision-making and be set 'right' prior to major decisions taking place regarding promotions, team deployment or any other resourcing activity. Figure 4.5 illustrates this.

The other aspects of strategic HRM (careers, training and development, and employee involvement) can be directly influenced by effective employee resourcing decision-making practices (detailed discussion on resourcing follows in Section 4.5). Indeed, it could be argued that they are best managed via carefully designed and managed resourcing procedures (Taylor, 2005). Such procedures would also benefit team formation and help in taking the employees' personal circumstances into account (Kochan *et al.*, 1986; Dainty *et al.*, 2002). This way, the impact of travel (inherent characteristic of working in the construction industry) could also be minimised. However, it is important to note that procedural change is a sufficient response to only some of the factors. Others, such as team spirit, must be managed via cultural changes as well as procedural guidelines. These variables are not so easy to deal with and require careful analysis and management if they are to be eliminated from the conflict area and moved among the compatibilities or their (negative) impact minimised successfully. This is particularly difficult to engender within the construction industry, where organisational traditions are ingrained in years of operational practice (see Chapter 1).

It is important to note that although 'training' and 'work–life balance' seem to appear on both, the conflicts and compatibilities, their scope within

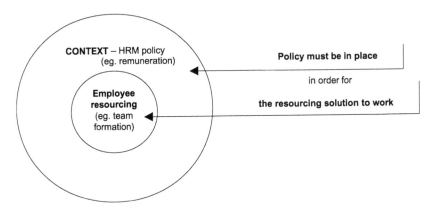

Figure 4.5 HRM policy vs employee resourcing practice.

each area is very different. Conflict related to training refers to the overall process: the organisation and management of training, including informal and on-the-job training activities, where the compatibilities related to the same factor focus solely on formal classroom training courses. Similarly, work–life balance conflicts comprise all aspects of employee work–life balance, such as long working hours, opportunities for flexible attendance patterns and teleworking, where in relation to the compatibilities it is considered only in the context of team formation.

Overall the data has indicated strength in the managers' approach and aim for good people management practice. This in itself supports a positive foundation for developing future opportunities for more organised strategic HRM practices. It is also a lever for lifting the barriers that hold the employees' trust back with regard to realising this aim. Despite this, it is important to note that only one of the factors in the compatibilities identified (work–life balance, a factor in team formation decision-making) is employee resourcing specific. Also, the recent growth and the high numbers of new employees brought into many organisations as a result of this, together with the data on conflicts, suggest that the current approach can no longer form the optimal method. A recurring theme is evident: lack of organisational policy/procedure leads to extensive confusion, uncertainty, disappointment and dissatisfaction. This suggests that the previously sufficient informal systems and individualistic management style are unlikely to provide support within the larger setting and deliver effective strategic HRM solutions; employees highlighted the requirement for more formal ways to manage the strategic HRM processes.

4.4 Strategic HRM or personnel management

To make an assessment of the nature of current people management practices within the industry, an analysis of the primary case study data in relation to Storey's 27-item checklist (1992) was carried out to explore whether the organisation operated within the personnel or strategic HRM paradigm. Table 4.5 reproduces Storey's checklist with an additional column that identifies the aspects of strategic HRM evident within the primary case study

Table 4.5 Primary case study organisational practice on Storey's 27-item checklist

Dimension	IR/personnel	Strategic HRM	Case study
Belief and assumptions			
1. Contract	Careful delineation of written contracts	Aim to go 'beyond contract'	✓
2. Rules	Importance of devising clear rules/mutuality	'Can do' outlook: impatience with rule	✓

3.	Guide to management action	Procedures/consistency control	'Business need'/ flexibility/commitment	✓
4.	Behaviour referent	Norms/custom and practice	Values/mission	
5.	Managerial task vis-à-vis labour	Monitoring	Nurturing	
6.	Nature of relations	Pluralist	Unitarist	
7.	Conflict	Institutionalised	De-emphasised	

Strategic aspects

8.	Key relations	Labour-management	Business–customer	✓
9.	Initiatives	Piecemeal	Integrated	
10.	Corporate plan	Marginal	Central	
11.	Speed of decisions	Slow	Fast	✓

Line management

12.	Management role	Transactional	Transformational leadership	
13.	Key managers	IR/personnel specialists	General/business/line managers	✓
14.	Communication	Indirect	Direct	
15.	Standardisation	High (e.g. 'parity' an issue)	Low (e.g. 'parity' not seen as relevant)	✓
16.	Prices management skills	Negotiation	Facilitation	

Key levers

17.	Selection	Separate, marginal task	Integrated, key task	
18.	Pay	Job evaluation: multiple, fixed grades	Performance-related: few if any grades	
19.	Conditions	Separately negotiated	Harmonisation	✓
20.	Labour-management	Collective bargaining contracts	Towards individual contracts	✓
21.	Thrust of relations with stewards	Regularised through facilities and training	Marginalised (with the exception of some bargaining for change models)	
22.	Job categories and grades	Many	Few	
23.	Communication	Restricted flow/indirect	Increased flow/direct	
24.	Job design	Division of labour	Teamwork	
25.	Conflict handling	Reach temporary truces	Manage climate and culture	
26.	Training and development	Controlled access to courses	Learning companies	✓
27.	Foci of attention for interventions	Personnel procedures	Wide-ranging cultural, structural and personnel strategies	

organisation with a tick. The discussion on the analysis is in sections under the headings extracted from this table: belief and assumptions, strategic aspects and line management, and key levers.

Belief and assumptions

It is clear from Table 4.5 that the primary case study organisation operates an overall personnel type approach to people management. However, aspects of strategic HRM are apparent in statements drawn from the analysis, such as [1] 'Aim to go "beyond contract"', [2] '"Can do" outlook' and drive on [3] '"business need"/flexibility/commitment'. The organisational strategy and values (discussed earlier) support this. In particular, the company's commitment to empowerment and training and development, together with the friendly culture, arguably provide a firm footing for a more strategic approach towards HRM.

Strategic aspects and line management

The organisation is also well positioned in relation to its line-management capabilities. Central to the concept of strategic HRM are [8] business–customer relations and [13] devolvement of key management activities to the operational managers, which are both crucial to the way the organisation is managed. The devolution of key management activities to operational line managers was a particular feature, a trend also identified by Druker and White (1995, 1996) as a characteristic of personnel/strategic HRM within the industry as a whole. Furthermore, the role of the HR specialists was found to be operational and reactive, also in support of Druker and White's (1995) findings. Perhaps due to the devolvement efforts, many line managers operated [9] independent HR systems within their divisions, and viewed HR specialists as intrusive. These concerns intensified when coupled with the production-oriented leadership style of some senior managers. Few managers showed a 'balanced' team management approach, which places maximum concern for both people and the production process. On the whole, the role of the HR specialists and line managers' focus on meeting the objectives of the construction process suggested the organisation operating what Purcell and Ahlstrand (1994) termed a 'paternalistic' approach to employment relations (Chapter 3).

Key levers

The organisational intention to [19] harmonise the terms and conditions of employment for all staff and [20] individualise the company employment relations framework supported the organisation's strategic HRM objectives. In addition, the [26] learning company culture and training and development policy and practices were found to form particularly powerful HR tools. These were managed and promoted through the company perform-ance appraisal system. Unfortunately, the appraisal system failed to deliver on the many well-intentioned schemes (this is discussed in more detail later). Moreover, the organisation operated a fairly reactive approach to [25] managing conflict. Managerial behaviour was focused on following estab-lished traditions and company practice, a key feature of the 'paternalistic' approach to employment relations alluded to the aforementioned. Task set-ting was motivated by the need to monitor staff/labour. In addition, [23] indirect communications and short-term operational view on the majority of the key levers: [17] selection, [18] pay, [21] thrust of relations with stewards, [22] job categories and grades, [24] job design, were common. This placed the organisation's view on the psychological contract construct towards the transactional end of the relational-transactional continuum (Chapter 3).

Summary

In summary, the analysis of the primary case study organisation's belief and assumptions, strategic aspects and line management provide supporting evi-dence for a strategic approach towards HRM. However, the [25] reactive approach to managing conflict, 'paternalistic' managerial behaviour focused on [4] following established traditions and company practice, [5] task set-ting motivated by the need to monitor staff and [9] piecemeal personnel initiatives, [10] marginal to the corporate plan reduced the organisation's projected potential for a strategic HRM type approach. In addition, the [23] indirect communications and short-term operational view on the majority of the key levers [17, 18, 21, 22, 24] place the organisation's view on the [12] psychological contract construct towards the transactional end of the relational-transactional continuum. Thus, a personnel type approach to people management remains dominant within the primary case study organisation.

Throughout all case study organisations the role of the HR specialists was operational and reactive. Perhaps due to the devolvement efforts, many line managers operated independent HR systems within their divisions, and viewed HR specialists as intrusive. These concerns intensified when coupled with the production-oriented management style. Few managers showed 'balanced' team management approach, which places maximum concern for both people and the production process. On the whole, the role of the HR specialists and line managers' focus on meeting the objectives of the con-struction process suggested what Purcell and Ahlstrand (1994) termed a 'paternalistic' approach to employment relations.

A key feature of the 'paternalistic' approach was a fairly reactive approach to managing conflict. Managerial behaviour was focused on the following established traditions and company practice. Task setting was motivated by the need to monitor staff/labour. In addition, indirect communications and short-term operational view on many HR functions, such as recruitment and selection, remuneration and job design, were common. This places the view on psychological contract construct towards the transactional end of the relational-transactional continuum (see Section 3.3).

However, in assessing the compatibility and conflicts between managerial and employee respondents in terms of their resourcing priorities, a total of 139 variables were identified. Five key themes emerged: team/project, employee involvement/communications, careers, learning and development and organisational planning. While a range of conflicts (and compatibilities) became evident, overall, the results suggested that with some attention there is significant potential for development towards strategic HRM style decision-making.

4.5 Learning and development

Learning and development was one of the key topics discussed within many of the research interviews. The literature suggests that construction organisations commitment to learning and development is limited due to the belief that it is a costly function, which potentially makes the company employees more attractive to competitors. In addition, it is reasonable to assume that the industry's macho culture and short-term focus on operational issues may prevent many managers from seeing the long-term benefits of organisational succession planning and individual career development. In light of this, it was somewhat surprising that the research revealed that a leading employer within the industry adopted what amounted to a highly sophisticated approach to training and development. They were found to actively encourage continuous development and facilitated self-responsibility and inter-organisational learning through the temporary organisational structures.

Both managerial and employee respondents within the primary case company felt that training and continuous development were encouraged and supported. This supports the company's People Statement on learning and development. Training towards professional qualifications and gaining chartered status as well as continuous professional development were high priorities within management ranks. They focused on promoting staff development through the appraisal system, which formed the formal means of discussing, identifying and recording employee training needs.

The appraisal interview provided an opportunity for discussing potential progression solutions and aided assessing individuals' current job performance, developing personal development plans and recording employees' aspirations and preferences. The system included both objective and subjective elements. The measurable (objective) aspect focused on evaluating performance and progression solutions and identifying related training and development needs. The subjective element sought to extract employee thoughts and satisfaction in relation to the interpersonal relationships within the team, department/wider organisation and the HR/operational processes. Full records were signed by all parties involved and progress followed up in six-monthly reviews and/or in the following year's annual appraisal as appropriate. Summaries of the individual training and development needs were collated within a bespoke database and distributed for divisional senior management approval. These were then brought together with the overall business plan to form the basis for wider organisational development plans, which the HR director used to assess and distribute budgets as necessary (the appraisal system is discussed in detail in Section 4.5.)

The formal training interventions supported by the organisation included training towards professional qualifications, such as corporate membership of the Chartered Institute of Building (CIOB) and Royal Institute of Chartered Surveyors (RICS), continuous professional development and day release part-time degree study at local universities.

More informal development mechanisms included mentoring and coaching, job shadowing, induction programmes, developing potential courses, encouragement of innovation and sharing of good practice. The mentoring and coaching schemes were used to provide a point of contact for both newcomers and managers rising through the organisation, from whom they can obtain informal careers advice, encouragement and support. This approach was also used to help instil the company values on all managers within the organisation. Job shadowing and induction programmes were introduced to support new recruits and recently promoted staff. This helped to familiarise new recruits with the company policy and practices. New senior managers were given the opportunity to job shadow an existing senior member of staff in order to facilitate their integration within the firm. The company had also sought to develop future potential in collaboration with a leading management college. Clusters of managers and other personnel identified for succession planning were invited to attend appropriate training courses. Bringing together the clusters of people from different areas of the business on this programme encouraged new practices and innovative approaches to be developed and their effective application throughout the organisation. Regular weekly meetings between senior managers and directors were used to encourage innovation and further sharing of good practice. New ideas and practices emerging from individual employees and project teams were evaluated and discussed in order to help to transfer good practice throughout the organisation.

Single/double-loop learning and adaptive/ transformational learning

Different forms of learning were discussed in Section 3.2. The organisational commitment to training and encouraging continuous development indicated commitment to single-loop learning. Short-term training interventions were used in response to performance issues and/or legislative changes and other environmental influences. Both managers and employees felt that the training towards professional qualifications and continuous professional development were high priorities within the organisation. The formal train-ing interventions were supported by a range of more informal development mechanisms, such as mentoring and coaching, job shadowing and induc-tion, encouraging innovation and sharing of good practice schemes. The informal activities, together with the continuous professional development, emphasised the role of longer term development via double-loop transform-ational learning. This suggests that the primary case study organisation's commitment to training and development in terms of the single/double-loop learning and adaptive/transformational learning follows the principles of an effective learning organisation. The company provides for and encourages structured training courses. In terms of staff development, their long-term strategy is to achieve fully qualified workforce. In addition, high importance is placed on continuous professional development. This mix involves short- and long-term solutions to continuous development of both the organisation and individual employees. The mentoring and coaching schemes help to ensure that the employee needs and preferences are inte-grated in the planning and delivery of learning and development activities. However, although learning and development were clearly integrated into the organisational culture, there was no evidence of active reflection (learn-ing to learn), the ultimate goal of a learning organisation.

The learning process

The organisation was found to successfully acquire new knowledge through multiple methods. Training courses provided basic information and updates on issues such as legislative change and thus helped to adapt operational processes to environmental change. Knowledge sharing was facilitated through more informal mechanisms such as mentoring and coaching. How-ever, knowledge utilisation (the third and fourth aspects of Nevis *et al.*'s (1995) and Huber's (1991) models (see Section 3.2) represented a weakness for the case study organisation. Many respondents discussed this to result from the dispersed and temporary organisational structures that the project-based environment dictates. Knowledge transfer was found to be difficult in forms other than individual employee knowledge carried forward from one project to another. The company had team meetings and project-end evalu-ations in an attempt to facilitate inter-project learning; however, limited use of information technology in recording the outcomes hindered wider transfer of the knowledge gained from these events.

The other element of the learning process, social construction that refers to the self-reflective process involved in transforming cognitive learning into abstract knowledge and the symbolic and political processes involved in learning, was found relatively effective at the level of the individual. The transformation of cognitive learning, which results from the know-ledge acquisition–sharing–utilisation process mentioned earlier, into abstract knowledge, was evident in the continuously increasing level of skill and competency staff hold. However, this failed to achieve the desired level of learning, mainly because of the difficulties in knowledge utilisation. Nevertheless, the symbolic and political processes related to social construction of learning strongly highlighted training and development as being the key to organisational success and individual advancement.

Systems thinking

The quantitative analytical measures taken in order to elicit the importance of learning and development in the strategic HRM decision-making from the interview and questionnaire data revealed several factors that reflect Senge's five 'component technologies' [1–5]:

1   Firstly, factors relating to organisational and HR planning emphasised the importance of long-term planning together with organisational flexibility and management of change. This reflects Senge's component technology 1 of the learning organisation: personal mastery. Flexibility was used to allow for continual clarification of the organisational focus and change management initiatives supported effective and timely implementation of the vision and mission.
2   The second component, mental models, was reflected in the organisational culture which was founded on trust, openness, partnering, empowerment (employee involvement) and individualistic management style. The company's choice in terms of operational and human resource management overall was to devolve many of the responsibilities to line management. It was the responsibility of divisional directors and senior managers to ensure that the operations run smoothly and the personnel involved in projects were looked after appropriately. Project-based personnel also had the remit and accountability for their particular elements of the work. Employee preferences were incorporated into project deployment decision-making through the close relationships departmental managers had with their staff. In addition, employees were closely involved in their personal development and career planning.
3   Thirdly, the learning and development and careers themes drew attention to organisational development and continuous improvement. Transparent progression opportunities and succession planning activities focused staff retention and achievement of organisational goals in the long-term. Career development, fast track progression and taking on trainees balanced this with extensive employee opportunities. These reflected the organisational commitment to building shared vision, Senge's component technology 3.
4   In terms of the importance of a team as the central unit in development (Component 4) the respondents highlighted the significance of good team spirit and relationships. The same key themes were found to be significant within the employee interviews and the analytic hierarchy

method questionnaire results also supported this. Good team relationships together with personal and professional development and gaining broad and/or specialist experience were ranked as the most important factors to be taken into account in strategic HRM decision-making.

5   Finally, the analysis on systems thinking (Senge's component 5) demonstrated very high managerial commitment to learning and development. The developmental philosophy was strongly rooted in the organisational culture and ethos, and learning activities were embedded in the daily operations. Learning and development was led by a strategy that provides clear direction and motivation to encouraging training and development at all levels and stages of projects and individual jobs. Line managers and HR personnel supported this view through their transparent commitment to promoting learning and development. This achieved high levels of staff satisfaction.

Although many aspects of the company practices clearly reflect a learning organisation culture, it is important to note that neither the interviewees nor the questionnaire respondents recognised this as an appropriate 'label' for their intended approach. The respondents' apparent unawareness of the terminology has a significant implication in delivering the espoused goals of the learning organisation within the construction sector. The literature indicates that many company programmes that develop learning organisations fail to deliver the desired results because the initiative is viewed as a destination, rather than an on-going, continuous process. Since the case study organisation clearly embraced the values and principles of the concept, without recognition of appropriate terminology, this shows their 'true' commitment to advanced learning and development; and further, to becoming a learning organisation. Despite an organisation achieving recognition for their approach to learning and development, they continually encourage further development and improvement.

Understanding the dimensions of organisational learning

In relation to the four central aspects of Nyhan *et al.*'s (2004) model of understanding the dimensions of organisational learning [1–4], evidence of advanced learning and development practice within the case study organisation supported the aforementioned contention of a learning organisation:

1   There was clearly a balance between formal structure and informal culture, as revealed by the third element of analysis which evaluated the compatibility and conflicts between the organisational priorities, project requirement and employee needs and preferences within the case study organisations' strategic HRM strategy, policy and practice. The organisational structure was mostly hierarchical with the learning and development strategy providing a clear direction for encouraging

developmental activities at all levels. However, at the same time, the organisational culture was informal; described as 'friendly, open and family orientated' with many of the strategic HRM responsibilities devolved to line management. Furthermore, the achievement of organisational goals is monitored and assessed via the company performance appraisal system, which also included elements focused on meeting employee needs.

2   By the very nature of construction work (individual projects custombuilt to client needs), work within the industry was seen as being varied and challenging. Employees also enjoyed the transparent progression opportunities.

3   These provide key opportunities for learning, which are supported by managers at all levels through the formal and informal learning and development mechanisms.

4   Finally, the company's collaboration with a leading management college together with their integrated approach to provision of NVQs, professional qualifications and informal learning and development demonstrated the partnership approach taken to incorporate all aspects of training and development particularly well.

The result of this type of staff development policy and practice has been that staff felt supported, empowered and were able to take advantage of the full range of opportunities available within the organisations. The succession planning benefits that this provides has meant that the organisation's key personnel were long-serving members of staff who have reached their positions through the promotion and development processes. The open approach also benefited the organisation in that newcomers were encouraged to bring in their fresh ideas. Together, these management development activities ensured a culture of mutuality within a spirit of continuous improvement that is paying dividends in terms of the organisation's performance. This suggests that the case study organisation attempted to understand and reconcile the opposing dimensions of organisational life along the two continuums in Nyhan *et al.*'s model and have achieved an inclusive 'both-and' approach. Conflict is evident but this is accepted and managed constructively in order to further continuous improvement.

Evidence of a 'Chaordic' learning organisation

Dainty and Raidén (2006) provide a detailed consideration of the second framework for analysis on learning organisation, the chaordic enterprise (see Section 3.2). As shown hereafter, their discussion is clear in that the case study organisation conforms to this model. Both, the characteristics of the environmental context within which the organisation operate and the company values and practice, reflect the key features of a chaordic enterprise. The central characteristics of a chaordic enterprise are discontinuous growth, organisational consciousness, connectivity, flexibility, continuous transformation and self-organisation. These are explored in relation to the primary case study organisation hereafter.

Discontinuous growth

The cycle of discontinuous growth is well documented in the construction environment (Loosemore *et al.*, 2003). Upward fluctuations in the economic markets are reflected in the sector in sharp increases in organisations' work-loads. At times of downturn, there is commonly a radical reduction in construction activity. Infrastructure and property development are often the first areas of economy to feel the impact of recession and in boom these are usually the last sectors to regain investment. The case study organisation had experienced this, the most recent example being in securing large public–private-partnership (PPP) contracts while their communications business (building mobile communications support stations) had also expanded at an unexpected rate. Short-term (reactive) and long-term (strategic) training and learning activities accommodated staff deployment to these areas. Thus, the organisation was able to take up the new opportunities available and sustain profitable existence.

Organisational consciousness

The organisational consciousness, which places importance on creating a collective vision as the driving force for change, is another characteristic featured strongly within the case study company. As discussed earlier, there was a general conformity to informal, friendly, family-oriented organisational culture. This was supported by strategy and policy, which provided clear direction for employee efforts. The combination resulted in culture of mutuality within a spirit of continuous improvement throughout the organisation. This has learning and development at the heart of the operations at all levels.

Connectivity

The third characteristic of a chaordic enterprise, connectivity, emphasises the nature of an organisation as a whole, and a part of a wider system. This is one area where the case study organisation lacks unity. The organisational structure divided the firm into regional units, which formed independent profit centres. These are seen to form a whole only at management ranks at higher levels and within senior professionals who may be allocated work in different parts of the company. Contractors are generally agreed to form parts of extensive and complex supply-chains, which include, on the one hand, the client and their advisory and investor connections and on the other hand, suppliers of materials and labour (Wild, 2002). In addition, there are direct connections with various other stakeholders, such as the government and professional bodies who influence contractors' operations. Indeed, construction projects are said to form extended virtual organisations or teams (Charoenngam *et al.*, 2004), which in turn again form a whole and a part.

Flexibility

Flexibility is as central to the construction industry/organisation as it is to the chaordic enterprise. As was discussed earlier, construction is a project-based industry within which individual projects are custom-built to client specifications. In addition, since the industry's output is largely non-transportable, construction organisations are required to set up temporary organisational structures at dispersed geographical locations. A large proportion of the work is carried out outdoors and so weather conditions may place restrictions on progress. Consequently, construction organisations rarely plan for change but react to it as situations arise. Continuous learning, as well as continuous transformation, is inherent in this environment. Learning is clearly both adaptive, in coping with the current conditions, and transformational, in devising new ways of working and organisational structures that accord with changing business needs.

Continuous transformation

The case study organisation's recent success in PPP and communications businesses reflects their ability to exploit the opportunities of continuous transformation. These developments followed from the slow down of civil engineering works. In line with the principles of this element of the chaordic enterprise, the organisation was able to initiate change very early on in decline and avoid steep fall in their overall workload. This required extensive training and development to refocus the company's market holding. Rebuilding the organisation from instability created a surprising competitive advantage as it generated unique management development opportunities and also extended the flow of knowledge from outside the organisation in recruiting significant numbers of new personnel to match the demand.

Self-organisation

Finally, a culture of mutuality within a spirit of continuous improvement was shared by all employees, and this directed thoughts and actions within the company. This benefited the organisation by giving it a clear agenda for self-organisation and self-development, although the 'traditional' culture of the industry hindered some significant developments in improving working methods or implementation of new initiatives/policy industry-wide. In fact, the strong organisational mindset found within the case study organisation in some respects undermined its espoused aim to improve and embrace change, and the operational practice to deliver according to well-established traditional ways. Thus, the learning opportunities from new staff and innovative ideas did not always take-off as hoped and frustrations arose as a result. Improvements tended, therefore, to be incremental in nature.

Summary

The analysis of the primary case data on the different elements of a learning organisation, Nyhan *et al.*'s (2004) model of understanding the dimensions of organisations learning and the theory of the 'chaordic' enterprise suggest that there are several areas of learning organisation that the company embraces. Most significantly this is true in understanding the different dimension to organisational learning, discontinuous growth, organisational consciousness, flexibility and continuous transformation. This supports a move forward from the situation described as common in the mid-1990s: low take up of the Investors in People initiative and poor commitment to learning and development. The learning organisation presents an advanced approach to learning and development within organisations with emphasis on self-responsibility and continuous improvement. The organisational view on human resource development strategy is encouraging in its longer term view and informal practices that incorporate individual employee needs into the decision-making process.

4.6 Employee resourcing

Chapter 3 discussed employee resourcing literature in the framework developed after Taylor (2005). This structure (see Table 4.6) is transferred here for the examination of the organisational practices. The primary case data opens up the dialogue with Figure 4.6 and then on each section under the headings extracted from the Table 4.6. Project case studies are used to illuminate the contextual and contingent nature of resourcing decision-making and highlight some of the difficulties research participants reflected upon. Much of the secondary case data is used to showcase promising practice identified within the companies. These 'boxes' provide a transparent link to the Strategic Employee Resourcing Framework (SERF) discussed in Chapter 5.

Human resource planning

Within the primary case study, human resource planning was managed at an organisation-wide level but with certain aspects being devolved to operational managers (see Figure 4.7 for a process chart). An overall strategic plan was put forward by the board of directors with targets for each division to achieve with regard to staff development and retention. Senior divisional

Table 4.6 Employee resourcing (after Taylor, 2005)

Strategic HRM objective	Strategic HRM activity	Tasks involved
Staffing	Human resource planning	Strategic human resource forecast – an input; development of a human resource plan – an output
	Recruitment and selection	Identification and analysis of recruitment needs; drawing of job descriptions and person specifications; advertisement of the vacancy; shortlisting candidates; selection process utilising appropriate selection techniques (i.e. interviewing, assessment centres, etc.); selection of the 'right' candidate; induction
	Team deployment	Formation and building of effective teams; deconstruction and redeployment of teams
	Exit	Redundancy, retirement, dismissal, voluntary exit
Performance	Performance management	Continuous evaluation and performance appraisal; feedback and reward
	Career management	Promotion; personal and professional development planning
HR admin.	Collection, storage and use of employee data	Utilisation of appropriate HR administration system, e.g. manual filing system or a computerised human resource information system (HRIS)
Change management	'Change agent'	Ensuring proper recognition is given to significance of change; management of business and strategic HRM processes via which organisational culture and structure continually evolve

managers then reconciled the targets against their resourcing requirements with a view of ensuring that appropriately qualified and skilled staff were available and that there was a constant supply of new staff into their division. In the short-term, this involved the formulation of a business plan and associated strategies to meet its objectives. It also included running 'what if' scenarios by notionally allocating staff to projects for which the division had bid in order to identify the possible gaps and how quickly they could be filled. More organised forecasting included numerical HRP on volumes of staff required for meeting the objectives of the business plan. HR specialists were consulted as to the employee development that supported the divisions' succession planning.

The outcome of the typical approach to HRP was that organisation could foresee gaps in their resources and HR capabilities in advance of projects coming on stream. Although this was effective in principle, it was managed

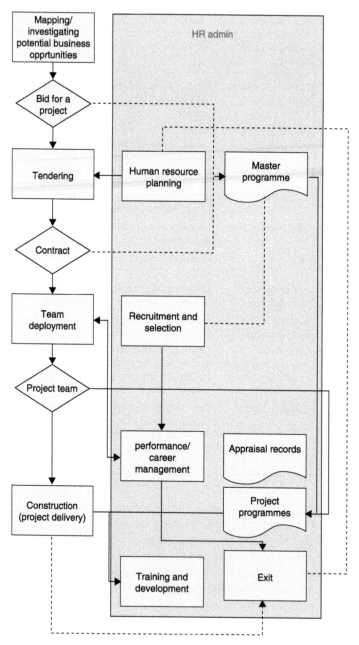

Figure 4.6 Overview of the employee resourcing activities (primary case study).

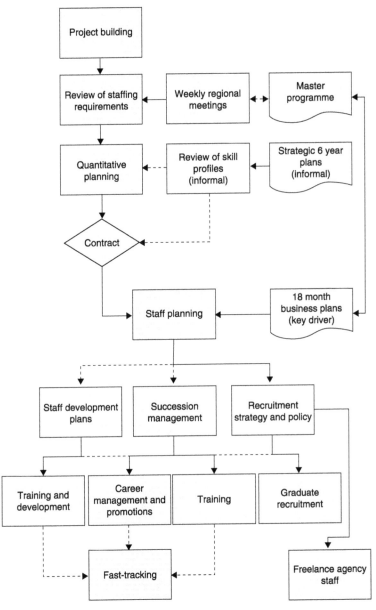

Figure 4.7 The human resource planning process (primary case study).

as a relatively isolated process in which the division explored its own staffing requirements largely out of context of the rest of the organisation. This may be caused by the devolution of responsibility for the management of employee resourcing activities to divisional directors and senior operational managers. Furthermore, the uncertain environment made effective HRP extremely challenging:

> The problem, especially with major projects, is that you never know which ones you are going to win. You may come to work one morning and the best thing you know is that there are two or three projects you are applying for. Staffing wise, you may either have enough people to do a 60 million pound job that you don't need, or suddenly overnight you may need all of them. So, determining how many people we need and how quickly we can get them is probably the biggest problem.
>
> (Construction director)

Nevertheless, the organisation's approach to strategic HRP suggests a likelihood of the senior management team integrating resourcing requirements with the business objectives.

Project case study: the distribution warehouse development

The distribution warehouse was a £16 million industrial development consisting of the construction of a freight terminal with two warehouses and an office base. It was a sudden, unexpected addition to the organisation's workload as the tender had been seen as highly unlikely to succeed. Thus, when the contract was won, the divisional managers did not have the resources available to staff the project. The construction director commented:

> I have been told we have a job starting 1st June, in 3 weeks time: 16 million in 22 weeks. That is going to need (a) a lot of people and (b) some really key people. And I sit here today and I actually don't know who they are.

The main difficulties in staffing the project stemmed from many of the organisation's key project managers and engineers being employed on the other two large projects being undertaken concurrently. The remote location of the project site also added additional pressures to the situation. The team members would have to lodge away for the project duration, a requirement that needed to be carefully considered. The long hours (due to the fast track nature of the contract) coupled with staying away from home Monday to Friday for 22 weeks, did not present the most attractive option for many members of staff. Thus, two younger engineers with no family commitments and an agent keen for promotional opportunities were approached as potential core personnel for the project. They all accepted the offer and soon were moved to prepare the site for the construction phase. The core team was supplemented with freelance staff and two permanent foremen. This brought together a 12-member strong team consisting of the contracts manager, project manager, project co-ordinator (freelance), works manager, senior engineer, four site engineers (two permanent and two freelance), a QS (freelance) and two general foremen. Five out of the 12 staff were interviewed for the project case study, including the contracts manager, project manager, two permanent engineers and one of the foremen.

During the early stages, the progress on the project was faster than expected. This allowed for the young engineers to undertake planned training courses in support of their professional qualifications and health and safety certificates. Unfortunately however, their performance appraisals were long overdue. This was due to the staffing changes and shortages near the completion of their previous project, which had resulted in all non-urgent activities being postponed. The realistic prospects for a constructive discussion with the new site agent only few weeks after the start of the project were naturally bleak.

This clearly illustrates the frustration unconnected and/or badly managed employee resourcing and other strategic HRM activities arise. The young engineers clearly showed aspiration for development and progression, but found little support in the organisational strategic HRM framework. Appraisal data were not included in the deployment decision-making, neither was it used to support the development of the engineers and other personnel on the project. Managerial actions were based solely on their subjective knowledge of their staff, and the employee voice was included only where individuals' had firmly expressed their views or actively showed interest towards a given project/path/option.

In summary, the staffing challenges illustrated by this case study were as follows:

Short-term planning

- lack of forward planning: the project's unexpected addition to the organisation's workload;
- managers unprepared and staff availability minimal: key personnel employed on other high profile projects;
- unattractiveness of the project due to the geographical location and fast track nature of the contract, which required long working hours.

Poor performance/career management (succession planning)

- long overdue performance appraisals due to previous project priorities/ commitments;
- inability of the new project manager to give constructive feedback on employee performance/progress at the start of the project;
- employee aspirations for development and progression not supported by organisational strategic HRM/employee resourcing procedures/ practices;

- isolation of the strategic HRM/employee resourcing decision-making processes;
- limited employee involvement in the process.

This project case study highlights the importance of consistent implementation of organisational policy and practice. The many areas of good practice evident within the primary case can be 'undone' by such circumstances. This constitutes a serious breach of the psychological contract and hence, has the potential to disengage the employee from otherwise fulfilling employment relationship. Effective planning is the key to managing the chaotic nature of short-term deployment demands and reducing the negative effects this may have on performance management and career development.

Promising practices in HRP

Succession planning

One innovative approach to HRP focused on succession planning. This involved the identification of people who showed director-level potential, who were subsequently placed on executive development programmes. In company A, the process was informal, managed by operational line managers with occasional support from HR specialists. In company B, HR specialists and departmental directors collaborated in identifying suitable candidates for their intensive management development programmes. Their selection criteria were drawn from business plans and current organisational capability charts. A named member of the HR team had the overall responsibility for overseeing and facilitating the process providing a single contact point for the directors, HR personnel and staff involved. This ensured effective integration of the organisational strategic and operational requirements of the business with a management team capable of providing the services and products that clients demand.

HRP schedule

Another successful HRP technique was found in company C, which operated a quarterly HRP schedule. Key managers met regularly to discuss the HR requirements for the following quarter in relation to the forthcoming workloads. Staff availability charts drawn from a resource management database were used as an information source for the meetings. This type of planning process was said to be particularly useful in identifying and balancing peaks and troughs in staffing requirements. Although, the quarterly schedule represents a relatively short-term outlook on staffing issues, the system helped the organisation to introduce structure to the process and reduce the uncertainty inherent within the industry's staffing practices (for a general discussion on the context and approach to people management in the construction industry see Chapter 1).

Graduate recruitment

Many organisations also placed great importance on graduate recruitment. Showing long-term commitment to developing graduates and offering them transparent progression opportunities were seen as key factors to successfully retaining the brightest candidates. Companies A and D, in particular, specialised in student and graduate recruitment as a long-term staffing strategy. This type of long-term approach to graduate recruitment illustrates effective integration of the strategic HRP and recruitment and selection activities integrated with learning and development.

Summer and industrial placements

Company E focused their efforts in establishing longer term commitment with summer and industrial placement students identified as being high-potential candidates for fast track development. They also paid particular attention to recruitment and selection at lower levels with the aim of recruiting good quality candidates and developing them through the organisation. At the time of the research interviews, the company was putting together an internal computerised HRP system. The system was described of having capabilities for availability and skills/experience scanning and mapping long-term staffing requirements. The company's approach to establishing early commitment and recruitment at lower levels emphasised the importance of the cultural management aspect of HRP.

Need to integrate HR-business planning

It is clear from the aforementioned that in many instances, HRP supported the organisational strategic intentions. Senior managers compiled staff development and retention plans, which directed the shorter term business plan actions. This practice and the specific techniques in use, such as 'what if' scenarios and numerical forecasting, suggest major implications. A HRIS would offer significant benefits to the current methods, especially the 'what if' scenario planning. Numerical forecasting can be carried out by simple spreadsheet applications. These findings show great potential for the improvement of the organisations' HRP practices. To tap into this potential for improvement, it is imperative that an element of planning is incorporated into the employee resourcing framework via a HRIS component.

Organisational planning was also identified as one of the key themes in relation to the factors the managerial respondents found important to be taken into account in their employee resourcing decision-making. Human resource planning was highlighted as a route to organisational flexibility and effective management of change via a culture founded on trust, openness, partnering, empowerment (employee involvement) and an individualistic management style. This supports the work of Laufer *et al.* (1999) who argue that effective planning is particularly important within the dynamic project-based sectors, in that it can help reduce uncertainty, introduce structure and create order and action. The findings also support the work of Smithers and Walker (2000), which emphasised the need for effective planning in order to decrease the chaotic nature of a project via HR-business planning integration and culture management. However, despite the recognised importance of planning and support for literature in this regard, the findings and results do not show the current approach as effective, but rather indicate broad areas for improvement. Accordingly, Table 4.7 summarises the requirements for improvement together with the current practice and importance of HRP.

Recruitment and selection

A recent rapid growth in the organisational workload had demanded a sharp increase in the recruitment of new staff at all levels. This had included bringing in key senior personnel to run major projects, project team members and a number of support staff, such as HR specialists, IT and administrative support workers. Despite extensive recruitment efforts, only around 50% of the staff recruited had been taken on as permanent employees, with the remaining shortfall being made up with temporary agency staff. The industry skills shortages were widely discussed as having a significant impact:

Table 4.7 Current human resource planning activities in the primary case study organisation, importance of the function and need for improvement

Current practice	Importance of the function	Need for improvement
• Supports organisational strategic intention • Techniques in use: 'what if' scenarios and numerical forecasting • Organisational planning and HRP key themes in resourcing decision-making	• Potential route to organisational flexibility and effective management of change • Can help reduce uncertainty, introduce structure, create order and action and decrease the chaotic nature of a project • HR-business planning integration • Cultural management	• Introduction of HRIS technology • Structured support for the current methods, especially 'what if' scenario planning • Transparent HR-business planning integration • Cultural management support • Effective delivery of the planning outcomes to operational decision-making

Currently there is a lot of work in the UK. There are lots of opportunities for people and the situation is getting to the stage where you have a continuing shortage of skilled people. It is a big problem. From a stage where people were glad to have a job were are now moving to the stage where you only need to have a quick look at the trades journals and there are numerous jobs advertised at the backs of those.

(Design co-ordinator)

Additional pressures on the recruitment and selection process (Figure 4.8) had also become apparent from the need to shift towards new market opportunities in the PFI and commercial building sectors following a parallel decline in infrastructure works over the past few years. Different management competencies were required for such positions, which were difficult to acquire. Moreover, the rapid changes in workloads led to increases in important recruitment activity short-term:

Recently, we got a contract ... which is a 40 million pound job. It was really sudden. We didn't expect to get it at all. And we had to recruit very quickly a senior person for it. So that was almost overnight recruitment ...

(Construction director)

Word-of-mouth recruitment and headhunting played a significant role in identifying new managers. The senior management team felt most comfortable using these techniques in ensuring that the new entrants have core interpersonal qualities such as a keenness to work as part of a team, assertiveness (but not aggression), ability to fit within the organisational culture and good communication skills. Technical competence, previous experience, personal skills and knowledge and personal ambition were also seen as important characteristics of the managers that were likely to take the business forward. One innovative approach was that senior managers and personnel staff sought to determine these qualities through the provision of scenarios, a selection interviewing technique known as 'behavioural interviewing'. Divisional managing directors monitored the process for senior positions, but lower level vacancies were filled at the discretion of line managers at a project level. This decision-making protocol proved to be effective, although many managers responsible for the recruitment and selection at project level had not been trained for the role:

I haven't had any formal interviewing training. I have been through booking schemes and trial and error. But I suspect I have not been allowed into this position and been promoted to do those interviews until I have demonstrated that I have a fair chance of getting the right type of people because of my background, I know the culture and I have been developed as a senior manager in line with the rest of the senior managers with a common focus, common ambition.

(Operational senior manager)

Senior managers and employees alike discussed the importance of getting the recruitment and selection decisions right if the culture, which had ensured the retention of many of the organisation's longest standing staff, was to be maintained. The HR director noted:

> We have always had it difficult to bring people in at the senior level, to bring them into the company culture. It is not an easy thing to do. People grow in our culture and receive the training, then they can develop within the company.

The influx of so many new staff had had a marked negative effect in that it had contributed to a dilution of the strong 'friendly, open and family-oriented' culture upon which the organisation had been founded. Many long-serving employees were used to 'open door' policy of communication, which became impossible to maintain in sections which had high increase in workload and thus volume of staff. In some instances, key managers had up to 150 employees reporting directly to them. Where employees were also used to informal contact with their managers dissatisfaction was apparent as again, the informal contact was difficult to retain. While managers attempted to deal with these issues sensitively, inevitably some friction arose at site level. Particularly younger newcomers, who had been recently educated, had quite different working ethos, style and personal needs to many members of staff who had served in the industry for years (see also analysis of employee needs and preferences in Section 4.3).

The operational senior managers believed that poor graduate recruitment and development were holding back the expansion and improvement of the industry's resources as a whole. The process of recruiting and retaining quality candidates was identified as becoming increasingly problematic in recent years. Fast track progression opportunities were offered to attract and retain suitable candidates. However, although this had allowed for fresh ideas to be brought into management decision-making, at the same time it had also resulted in resentment from some existing staff members, dilution of the organisational culture and even to certain individuals suffering from stress-related problems. A senior contracts manager referred to one of his employees on such a scheme:

> The question is how far can you stretch a person . . . we have had, luckily only a small number of incidents with individuals suffering from stress related illnesses. But once you have somebody it is horrific. I have got an individual working on one job who hit exactly that at Christmas. We basically sat him down and he had time off. We put a lot of time into giving him support, bringing him back. We put him into a job role that he was comfortable with. Now, I did his appraisal last week and it would appear that he is back on strain now. I believe we might have pushed him too hard again . . . Somebody who is on the surface capable

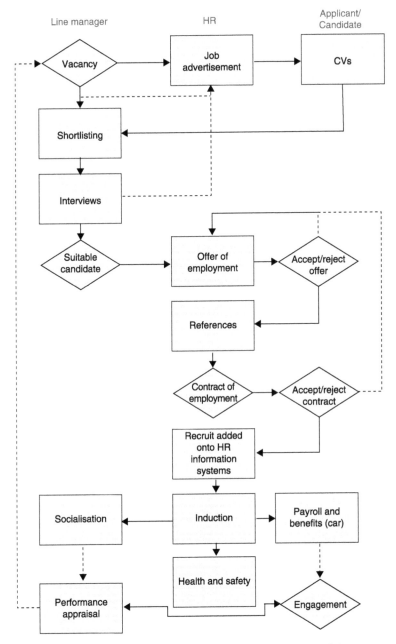

Figure 4.8 The recruitment and selection process (primary case study).

of taking anything really and then in front of you snaps. That is one of my key issues with human resources . . .

Promising practice in recruitment and selection: considering corporate image, internal recruitment, job descriptions and centralised HR services

Corporate image

Company A prioritised the image of the company as an employer in their recruitment and selection procedures. The suitability of the company's value system to that of the potential candidate's was said to be one of the main determinants of success. Thus, they actively promoted the company culture within the recruitment process and sought to deliver positive messages using media that aligned with the company's values. Apart from the professional publications, this also involved school visits and university milkrounds. This type of approach clearly targeted their recruitment and selection efforts at a population that potentially fits the organisational culture.

Internal recruitment

Company C focused on internal recruitment and promotion. All vacancies were advertised on the company Intranet and the staff was regularly encouraged to access the available information. This provided transparent progression and development opportunities, which encouraged long-term commitment to the company.

Job descriptions

Company D made extensive use of job descriptions in their recruitment and selection process, both internally and externally. The job descriptions included two levels of job-specific roles and responsibilities. Table 4.8 illustrates this. What was described 'an entry level' included the minimum requirements, capabilities and competencies to carry out the duties involved in the post. These were used to assess potential candidates' suitability for taking up a post. The other level specified the skills and requirements for an advanced understanding of the tasks involved in the post. This specified superior performance. The level was termed 'a promotion level' as usually meeting the requirements of this level led to the post holder taking on additional responsibilities and being promoted or transferring to a different role via lateral move. The company policy was to recruit candidates, when ever possible, at the entry level and progress them through to the exit level. This had been proven successful in building longer term commitment and ensuring staff have the right skills and qualities required for the post. A comprehensive induction programme, which included introduction to the company policies and procedures during the first day of employment, was also said to aid the process.

Centralised HR services

Company E built on a centralised recruitment and selection service managed by the HR department. The responsibilities were clearly divided in two: graduate recruitment and executive recruitment (managers, qualified professionals). Operative labour recruitment was devolved to line management. This provided staff and managers within the organisation with a single point of contact. The company had also initiated an overseas recruitment programme. This was said to successfully cover for shorter term (up to three years) requirements and occasionally longer term requirement too. Similarly to company A's graduate recruitment service, company E's centralised recruitment and selection activities provided the personnel involved with a benchmark for the general standard and quality of the potential candidates. In addition, the centralised service helped the HR department to ensure that the widest possible pool of potential candidates was reached by the recruitment efforts.

Graduate recruitment

As within HRP, graduate recruitment was also paid particular attention to in terms of recruitment and selection. Many organisations committed considerable resources to it. In company A, selection was based on the personal characteristics of each potential candidate. The company paid for the candidates to attend interviews and assessment-centred activities. Departmental directors were involved in graduate open days to show strategic high level commitment. The process was managed centrally through the HR department. This was said to be a cost effective and efficient method, and also to provide the personnel involved with a benchmark for the general standard and quality of the potential candidates.

Isolated strategy and informal methods

The organisational recruitment and selection practices were often isolated from HRP. Informality was identified as an inherent and integral aspect of the process, which had resulted in highly fragmented systems of operation within local divisions/projects. Vacancy information was not available

Table 4.8 Job description outline

Entry requirements	Promotion criteria
• Basic IT skills (including MS office, e-mail, internet) • Teamworking skills • Ability to work independently under the supervision of a team leader/project manager	• Advanced IT skills (including MS office, e-mail, internet) and an understanding of CAD • Team leadership skills • Initiative, innovation and responsibility

company-wide and thus employees were not always aware of potential opportunities within the group. Equally, the widest possible pool of candidates was not attracted outside the organisation due to the methods in favour: word-of-mouth and headhunting. However, despite the recent rapidly increasing demands, managers found the system effective.

The informality of the organisations' recruitment and selection practice strengthens the evidence provided by the industry literature, which highlights personal introductions common and important source of recruitment at all levels (Druker and White, 1996) and selection methods restricted to interviews and assessment centres (Langford *et al.*, 1995; Druker and White, 1996; Loosemore *et al.*, 2003). This undermines the vital importance of effective recruitment and selection process, which can be achieved via reconciliation of the HRP outcomes with the shorter term operational conditions. The transparent link between HRP and recruitment and selection processes can help ensure appropriate supply of skilled staff that positively contributes towards the achievement of the business objectives (Larraine and Cornelius, 2001). Further, the type of benchmarking exercise inherent in company E's centralised recruitment and selection activities helped the organisation ensure that the widest possible pool of potential candidates was reached by the recruitment efforts. Larraine and Cornelius (2001) suggested this as one of the key features of effective recruitment and selection practice. Accordingly, it is suggested that the recruitment and selection processes are integrated within the HRP aspect of the staffing function. Table 4.9 summarises the elements required.

Team deployment

Team formation and deployment was considered to be the most important of all the aspects considered under employee resourcing:

> The real issue in construction is whether you can form good teams or not. This makes the difference between success and failure. Actually half of your long-term success is in the strength of your team.
>
> (Senior operational manager)

Table 4.9 Recruitment and selection practice, importance and areas for improvement

Current practice	Importance of the function	Need for improvement
• Isolated from HRP • Informal • Fragmented • No vacancy information available group-wide • Word-of-mouth and headhunting prominent methods	• Reconciliation of HRP outcomes with short-term operational conditions • Ensure appropriate supply of skilled staff to the organisation • Contributes to the achievement of business objectives	• Introduction of structured methods • Transparent HRP-recruitment and selection integration • Benchmarking

Nevertheless, the process (see Figure 4.9) was also considered extremely challenging to manage effectively due to the short-term time scales that apply to most construction projects. The need to select, form and deploy a team rapidly placed considerable strains on the efficacy of the processes currently in place. Staffing a project with entirely new personnel was considered too risky, which had resulted in people with known abilities being taken from existing projects, even where this could cause problems elsewhere. However, the dynamic nature of the staffing situation had inevitably led to a breakdown in this principle as operational needs overtook strategic objectives. Some major projects had ended up being staffed by teams of entirely new staff with little knowledge of the organisation and its operating procedures or even by members external to the organisation. Selecting staff suitable for working with particular clients was therefore rendered extremely problematic and often led to teams having to be reformed during a project when they failed to perform as required.

The selection criteria for finding the suitable key people to head a project were almost unanimously stated being based upon, in order of priority:

1 Availability
2 Previous experience (ability)
3 Client preferences
4 Individual's need for a particular job to gain experience or training
5 Individual's personal aspirations (including their career management/development needs)
6 The ability to devolve responsibilities (e.g. to develop and give experience to trainees on a project)

These priorities demonstrate the industry's tendency to focus on meeting immediate organisational/project needs, therefore placing employees' preferences and aspirations well down the priority list:

First and foremost we have got to safeguard our interests, obviously to make profit, and to satisfy the client.

(Operational senior manager)

Staff appraisal records were rarely considered in the decision-making process, with subjective senior management decisions being relied upon in the majority of deployment decisions. This relied upon senior managers' abilities to fully understand the capabilities of their staff, a task that was increasingly difficult given the rapid intake (and turnover) of staff in the current competitive labour market. Nevertheless, a senior estimator explained:

We probably do it from a gut feel really. I know the company, I have worked here for 27 years. You know people. There are newcomers, but I know the key players. It is all in here [pointing to his head] . . .

This intuitive process forms another example of isolated systems.

In addition to selecting the key personnel to head a project, ensuring a balance between the team members' strengths and weaknesses and their willingness to work together for a common aim was considered crucial.

It is getting the balance right, the team members are not identical. This guy may be quite good at this, and the other may be weaker at that. It is like a piece of jigsaw; if it fits well into the situation with this guy then between them they can be quite strong.

(Operations director)

However, weaker team members who did not necessarily complement other managers were still placed into teams once they had been released from their previous projects, regardless of their personal requirements of their new roles. Again, the need to resource project teams rapidly and a lack of information to inform the process had led to a strong likelihood that inappropriate decisions would be made. Indeed, such practices appear to render the practicality of the principles of structured forms of team formation suggested by Belbin (1991, 2000) and others as highly questionable.

The consequences of adopting such an informal and reactive approach resulted in a substantial loss of valuable knowledge. This was brought into sharp focus by the unfortunate and untimely death of the chief surveyor during the course of the research. In his interview, he made his exceptionally considerate management style very clear by referring to numerous examples of where a realistic balance between the business and individual employee needs had led to a successful long-term outcome. He explained the importance of mutual trust and respect:

People are not only coming to work to earn a good salary, they also want to satisfy their own personal aspirations in terms of the responsibility they have got and they want to feel a part of a management team. And I do think that is one of my jobs to try and make people feel they are part of a team ... to give them jobs which challenge them and secondly to create a climate and a perception in the company that they can go on ... I like to think I am flexible in this sort of position, I measure people's performance, not necessarily in the hours they work but by whether they can do the job. And if they have got any problems that is fair enough with me. I empower my people ... Well firstly, if they are involved in the decision-making process, if you get them to buy into it and they go away thinking they have been involved in it, they will want to implement it ... Secondly, I am a firm believer in that you work but you have got life outside of work. And what we are trying is avoid one person who lives on the east of the country travelling over to the west of the country, and the guy who is on the west of the country travelling over to the east. It is expensive in terms of operating the job, but more important the person who is doing the job is tired when they get there ... But, if somebody has worked away for some time, they know that I am not going to send them away again. It all comes down to trying to operate a fair system really. And have regard for people's own personal circumstances.

Clearly this manager personally espoused a strategic HRM approach in relation to team deployment, empowerment (employee involvement), performance and career management and work–life balance. Unfortunately, his approach reflected the informal culture of the organisation insofar as he had no records of his decisions and techniques. Hence, much of his valuable contribution was lost following the accident. The following quote he made during the research interview only three months prior to his untimely death emphasises the importance of effective recording mechanisms for continuity of practice:

There are no records to say that we have had so and so working away two years ago and he was out there for 9 months. And I don't think it is necessary to have a record of that. You might say 'well, if I leave or die or something like that who is going to remember it?' It depends on how seriously the other, my successor, would take that style of management ...

Team building was considered vital to the success of a project. Various levels and forms of exercises were found to be in use. Larger more complex projects involving vast numbers of new staff included cultural integration exercises to provide focus and feeling of belonging, whereas smaller teams relied on informal social events to foster team synergy. A team briefing structure was said to cascade down throughout the organisation. Divisional directors briefed their senior management teams, who then conducted similar meetings within their respective department heads. The department heads delivered the information to their project managers and senior surveyors and estimators. These personnel then met with their site-based teams. Inevitably, longer term projects tended to develop their own team sub-cultures, partially as a result of the considered effort to integrate people within them to work better together. This had led to problems when it came to breaking up such teams, as managers found it difficult to readjust into new team sub-cultures.

Somewhat surprisingly, considering the recognised importance of the team formation and team building in contributing to the overall success of the organisation, team effectiveness was not measured in any structured way. It was only considered where, for example, a team did not work well together. In such a case, making everybody aware of what is required of them and explaining why particular decisions were made was found an effective route to problem solving. If this was unsuccessful, some of the project staff tended to be re-deployed. This effectively defeated the original team selection process and hindered learning and knowledge sharing.

Case study: PFI schools project

The PFI schools case study consisted of a sub-section of a larger project programme. The overall programme included the construction of ten schools within a single region. This was divided into four sections, of which this case study was the first to commence (Figure 4.10).

The overall project programme was overseen by a contracts manager, who reported to the divisional construction director. A matrix management structure applied to the support functions, where, for example, the project surveying staff (QSs) were managed by the divisional chief surveyor rather than the contracts manager. Each section of the programme had a project manager. The project managers managed the section's production teams for the one, two, three or four sub-projects involved. Each sub-project had a manager or an agent, who was responsible for the supervisory staff (foremen, engineers, trainees) and directly employed labour. Most of the operatives were labour only sub-contractors.

Focusing on Section 1 of the overall programme, the PFI schools case study, Figure 4.11 provides a detailed structure of its sub-projects and appropriate support staff.

The Section 1 staff were all new to the organisation, apart from the sub-project two site agent (shaded in Figure 4.11). He was one of the longest serving members of the organisation with 35+ years experience in the region. The others had been specifically recruited for the project in question, this including the contracts manager.

All project personnel shown in the Figure 4.11, as well as the contracts manager, were interviewed as part of the study.

Projects within Sections 2 and 3 of the programme had significantly high

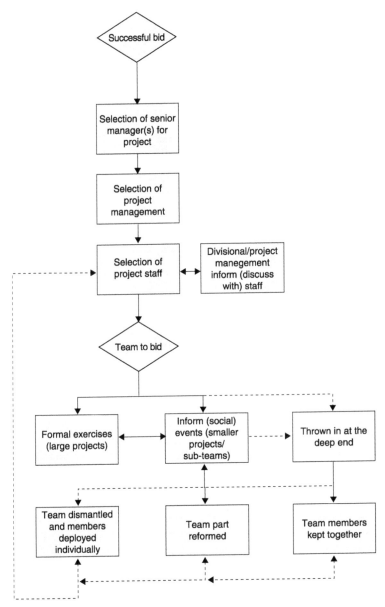

Figure 4.9 Team deployment (primary case study).

proportions of existing personnel supplemented with a few freelance and/or newly recruited staff. Section 4 had a relatively balanced mix of company staff and freelance support.

This type of staffing strategy resulted in multiple challenges throughout the overall programme duration. Initially, the contracts manager, being new,

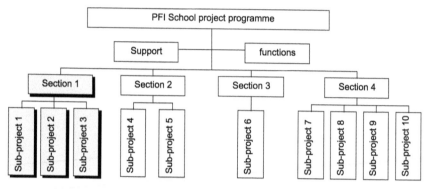

Figure 4.10 PFI school project programme.

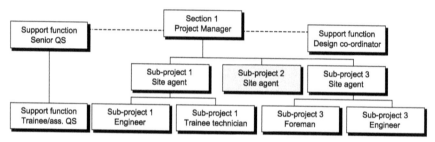

Figure 4.11 PFI schools case study project management structure.

required extensive resourcing guidance from his longer serving colleagues. As alluded to the aforementioned, the organisation had no written policy or procedures for project allocation/team deployment. Regular contracts managers meetings and the divisional directors' tentative human resource plans formed the core of the staffing activity. Thus, the newly recruited contracts manager was forced to rely on his colleagues as to judgements on the suitability of his resourcing decisions in relation to the company practice. Similarly, in recruiting new personnel, he had to actively seek advice so as to be able to follow established unwritten company practice.

At the site level, there was extreme pressure to meet expected performance levels. As alluded to the aforementioned, the case study section of the project programme (Section 1) was the first section to commence. All of the sub-projects were very similar in nature, and so the Section 1 was considered as an exemplar. Other sections and sub-projects could learn from the experience gained during the early planning and construction phases on the section and thus potentially achieve improved levels of performance. With this in mind, it would seem rather risky to allocate all new staff on the exemplar from which others could learn. Nevertheless, the section's 10 new recruits were trusted to undertake the work with the guidance of an older, very long serving site agent. On the positive side, this allowed for fresh ideas and working methods to be brought into the project and via that possibly to the organisation too, and useful lessons from external expertise to be learnt. However, this was at the price of unproven working relationships and hence, reduced team synergy. Many respondents on the project questioned why the later sub-projects (within Sections 2 and 3) had teams consisting of mainly existing employees, and suggested that a better balance between new and existing staff would have been beneficial on the 'exemplar' section.

In summary, the staffing problems highlighted in this case study were as follows:

Programme staff allocation

- the contracts manager in charge of the programme was new to the organisation;
- no written policy/procedure on recruitment and selection or team deployment;
- extensive team deployment and recruitment and selection related guidance required from longer serving colleagues;
- reliance on the appropriateness of the colleagues' resourcing decision suggestions in learning the organisational practices/culture;
- the ratio between new and existing project staff at an imbalance within the four sections of the programme: Section 1 mainly new personnel, Sections 2 and 3 significantly high proportions of existing staff, Section 4 relatively balanced mix.

Section 1 case study project team deployment

- 'exemplar' section within the programme: extreme pressure to meet expected performance levels and deliver learning from the planning and early construction phases to the subsequent sections of the programme;
- the project staffed with personnel new to the organisation;
- difficulties in team building due to unproven working relationships and 'power struggles';
- reduced team synergy.

Promising practice in team deployment: a holistic approach

Team/management development

Senior managers in company A emphasised the importance of effective team deployment to their success:

> Clearly the project is important, it is fundamental to what we actually do. But we need to build the projects through the people

that are working on them, rather than the focus of attention being the project outcome itself . . .

This implied a need for change in managerial behaviour. At the time of the interviews, the overall responsibility for team deployment was devolved to line management. This had resulted in short-term, reactive practices, where existing teams moved from project to project largely together and any available people were deployed to fill immediate staffing needs. Management development programmes and employee involvement had been adopted to facilitate change. Employees were asked to voice their preferences, and, where possible, they were deployed accordingly. Company-wide team development initiatives included site-based football, darts and cricket team tournaments. HR information and help centre liaised with managers offering advice on any potential issues to them and members of staff.

Company E encouraged employee development via national high profile project opportunities. This was managed in HR–line management collaboration, with operational directors having the 'final say' on the basis of information provided by the HR staff.

The HR–line management collaboration in organising and managing the developmental activities ensured both the organisation's operational requirements and strategic HRM objectives were incorporated into the process. Employee involvement as a facilitator for change also integrated the individual employees' personal needs and preferences into the process. This forms a comprehensive methodology for long-term organisational and employee development.

National and regional divisions

Company B had recently restructured its operations to form two separate divisions: regional and national businesses. Within the regional businesses line managers took day-to-day responsibility for team deployment and other strategic HRM issues. HR personnel managed the national business deployment via a central resource bank, which members of staff could voluntarily agree to join. The central resource bank consisted of an employee skills database, which recorded the members' experience and qualifications. Use of such a database helped the company to ensure that only personnel willing or keen to travel were deployed to projects beyond the regional boundaries. This was useful for managing work–life balance issues.

Resource management database

Company C also utilised a resource management database to inform their team deployment activities. Their system catalogued all employees' job titles, their previous experience and projects, line manager, home address, etc. Availability charts provided a basis for decision-making, which was focused on identifying the required skills and competencies for the upcoming projects and selecting appropriate personnel to fill the vacant posts. Secondary criteria included appraisal records, career development needs, location, salary package and clients. All this information was held within the resource management database, and thus easily accessible. The system helped to incorporate a comprehensive range of factors into the managerial decision-making.

Holistic approach

Company D operated a particularly effective approach to the team deployment process. Four main sources of information were drawn together to form a comprehensive picture of a potential team members' suitability for a project:

- technical competence (from job descriptions, experience summary sheets and chartered status records);
- personal development assessment (to support technical competence evaluation and provide information on personal aspirations, needs and preferences);
- personal relationships (line managers' subjective knowledge on how the employee works with other people/as part of a team);
- time (employee availability, current project/commitments, potential disturbance of a move mid-project).

Once the team had been deployed, the company encouraged them to socialise and develop their working relationships early on. The clients, designers and sub-contractors were also invited to the events to encourage good relationships within the team, which extended beyond the contractors' staff including all parties involved in the design, construction and use of the project/its outcomes. A project team was recognised as being

> three-dimensional with the client up on top, the contractor in the middle, and subcontractors at the bottom . . . what we try and do is get all those levels to build relationships together . . .

So-called 'health checks' were carried out every two weeks to ensure progress was satisfactory according to the schedules and staff morale high. The most efficient utilisation of team members' knowledge was said to form the principal motive behind the dismantling and redeployment of the teams. In general, part of the team was moved to a new project with selected personnel also being disbanded to other locations. This allowed for effective application of knowledge from the project and its team members experience of other sites. Regional functional managers had been introduced to form a communication link between site and office staff. Their role had also been proven effective as an alternative for direct line management contact regarding appraisal interviews, career management discussions and grievance procedures.

Reactive problem solving on issues of strategic importance

In relation to project staffing, the research findings clearly indicate that team formation and deployment are the most important of all the resourcing functions. This supports Druker *et al.* (1996), Yankov and Kleiner (2001), Walker (1996), Olomolaiye *et al.* (1998), Spatz (1999, 2000) and Goldberg (2003). Despite this, much of the task was managed on an *ad hoc* basis, and in response to immediate project start requirements or problems identified within existing projects. Many problems stemmed from the tendency to move personnel with known abilities around according to the staffing requirements of newly acquired projects. This was often known to cause problems within the projects from which key personnel had been removed, but the approach was still frequently used so as to minimise risk and conflict within new projects. This highlights the industry's short-term outlook towards project allocation (and HRM overall) and is further emphasised by examination of the project leader selection criteria. Alarmingly, the respondents stated that 'availability' was the number one factor they considered in allocating staff to a project. This was followed by two other variables that focused on meeting the organisational/project requirements: the potential team member's experience and client preferences. Only after these were the individual employees' needs and preferences considered. This is a clear inadequacy in the organisation's current resourcing practices. Yet, in addition to the poor selection criteria prioritisation, in terms of a decision-making process, the current practice relied solely upon the senior managers' understanding of their employees' capabilities and needs. No readily accessible database of information was available to support this intuitive process, and so decisions were based on the managers' implicit knowledge of their staff. This isolated the process from the other employee resourcing functions that could have been usefully employed to support the team deployment activities.

The ad hoc and subjective nature of the deployment decision-making demanded extensive flexibility from the employees. This was mainly in terms of functional and geographical flexibility, although occasional requirements for temporal flexibility were also evident (see Chapter 3). Managers attempted to compensate informally for this. For example, they promoted earlier finishing times on a Friday if project was progressing as planned and/or additional days off were possible. No formal means of recognition for the employee flexibility were in place. Many suggested that flexibility was an integral aspect of the nature of work within the industry, and thus a 'requirement' rather than an act of commitment or loyalty.

The organisation clearly operated in a manner akin to Atkinson's (1981) 'flexible firm' model (Section 3.3). Emphasis was placed on the core group and agency workers and sub-contracting used extensively. Evaluation of the team deployment activities in relation to Volberda's routine, adaptive and strategic categories (Section 3.3); however, revealed a rather reactive approach towards managing flexibility. At a strategic level little external flexibility was evident. This extended to the take-up of internal strategic flexibility which was slow. For example, strategies were developed over time in reaction to changes in the environment rather than dismantled and radically changed. Of the adaptive flexibilities, multifunctional teams were naturally common due to the inherent nature of construction project teams, and managerial roles frequently changed according to operational requirements. The use of temporary labour was widespread, but the use of other forms of internal/external operational flexibility as a planned activity was less apparent. This highlights the need for a more coherent approach to managing flexibility to be adopted. Thus, facilities for balancing the requirements for flexibility between the organisation/project and employees are essential aspects of a more appropriate employee resourcing framework.

The isolation of the team deployment activities from the other employee resourcing functions undermined their potential to support the managerial decision-making and use of flexibility as an intended strategy. For example, HRP could provide overall staffing forecasts (Turner, 2002) and a performance management system gather data on employees' skills and preferences (Nesan and Holt, 1999). A HRIS would conveniently facilitate the collation of this data and make it readily available to support the team deployment decision-making (Tansley *et al.*, 2001). The current one-way approach towards flexibility could be balanced via comprehensive recording and utilisation of organisational, project and employee data (Broderick and Boudreau, 1992; Tansley *et al.*, 2001). Taking employee views and preferences into account in team deployment decision-making would show organisational commitment to their employees' needs and introduce flexibility that benefits them as well as the organisation (Mabey *et al.*, 1998; Taylor, 2002b). The primary case study organisation's fragmented staffing practice draws attention to the need for radical process improvement and integration if benefit from effective employee resourcing is to be realised. In addition to the areas discussed earlier, this must include methods for measuring team performance and effectiveness to address the fundamental weakness in team building and guidelines for carefully managing exits from the organisation. Process integration in relation to team deployment and the other staffing functions (such as HRP and recruitment and selection) should form central components of the more appropriate resourcing framework. Table 4.10 summarises this.

Table 4.10 Team deployment practice, importance and areas for improvement

Current practice	Importance of the function	Need for improvement
• Ad hoc, reactive to immediate project start requirements or problems • Fragmented, isolated from other resourcing activities • Staff availability, number one selection criteria in team formation, employee needs and preferences neglected • Reliance on managers' subjective knowledge • Flexibility managed informally • Team performance measurement weak	• The key to competitive advantage • Most important of all resourcing activities • Powerful motivator • Supports the achievement of the project/organisational goals leading to improved productivity • The intangibles of human interaction separate average performance from outstanding execution	• Longer term planning • Integration with other resourcing activities • Broad range of variables to be incorporated into decision-making • Recognition for employee needs and preferences • Balance between organisational, project and employee factors • Introduction of HRIS • Coherent approach to managing flexibility • Structured team performance and effectiveness measurement framework

Exit

None of the interviewees referred to any issues with involuntary forms of exit, such as redundancies or retirement. Indeed, as discusses earlier, many organisations were actively recruiting new staff for different areas of business. With regard to voluntary exit, salary and reward, more broadly, were cited as the main reasons for people leaving construction organisations. As a result, more individualised reward mechanisms, including a bonus system, had been initiated to prevent further problems from arising. A substantial investment in training, together with the maintenance of an informal and friendly organisational culture, was considered the organisation's strongest retention factors and priorities for the primary case study organisation. Exit interviews were carried out in most cases, with company E having particularly effective procedure in place.

Promising practice on exit interviews
Company A carried out exit questionnaires and analysed these thoroughly. Their human resource manager commented

> Unfortunately we end up spending a lot of time with exits/terminations, which creates a vicious circle ... HR function should be

about recruitment and retention. Currently we are dealing with the wrong end of the business and don't get an opportunity to input to line management as to the suitability of the people . . .

Hence, the main problems with staff turnover were said to stem from poor recruitment decisions. Management development was identified as a potential solution to improving their recruitment and selection, training and development, and people management skills.

At company E, exits were managed in HR–line management collaboration. Their procedure for dealing with voluntary exits is presented in Figure 4.12. This procedure was carried out with anyone indicating an intention to leave, including managers, professional staff, graduates, industrial placement students, etc. It was said to give the company a chance to persuade a person to stay and correct any issues that might exist within their workplace. This formed a useful addition to the company grievance procedure, which some employees had felt uncomfortable dealing with.

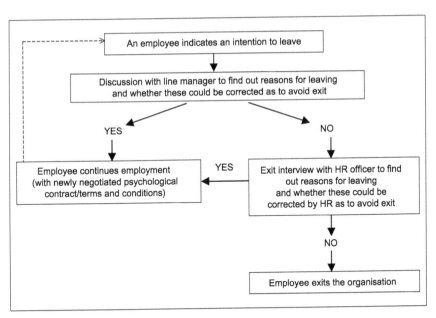

Figure 4.12 Company E voluntary exit procedure.

Macafee (2007) adds that exit interviewees should have a direct link to HRM at strategic level. Parallels with other forms of data gartering on employee satisfaction, such as annual employee engagement surveys, present an opportunity to gain valuable insights and enable comparisons between current and former employees. This can help in the development of bespoke retention strategies for different groups of employees. Also, where exit is handled sensitively, alumni can become useful brand ambassadors who will speak highly of their former employer and thus protect the 'employer brand'. As was noted by some of the employees in the primary case study organisation, one career development tactic is to leave with a view to coming back in the future. In such circumstances, whether this tactic is encouraged by the employer or initiated by the employee(s), it is important that the parties part in good terms. Indeed, one of the secondary case respondents (a Graduate Development Officer) kept contact with all place-ment students and former employees in order to retain a company associate network.

Careers

Recent organisational growth had resulted in many staff being promoted rapidly if they demonstrated appropriate leadership abilities. This has pro-vided ambitious individuals the opportunity to further their careers at a fast pace and realise their aspirations far more quickly than would have been possible during a stable or downturn period. However, there were concerns as to the viability of such practices in the future, as the organisation will not be able to fulfil the employees' raised aspirations and expectations which was likely to have detrimental effect on staff turnover. The following quote from a project manager illustrates the following concern:

> I am a project manager and I have said to him [the manager] I want to be a senior project manager, but we only have two in the company. Now whether there is a place for a third . . . You have to look at it from their point of view and be realistic. But then you look at it from my point of view and I couldn't care less how they react. I am good at the job and I can do the job and I need an opportunity to show them I can do the job. If they don't give me the opportunity, plenty of others. I mean, people ring in every week, we are being constantly headhunted . . .

On the other hand, the high level of expectation on both new and existing managers had resulted in certain individuals suffering from stress-related problems, as already discussed in relation to graduate development. Con-sequently, these individuals were said to require extensive periods of time off work and careful reintegration once they returned. Thus, senior managers were faced with the negative consequences of stress-related illnesses and the associated costs of time taken off sick, reduced productivity and morale, possible loss of valuable members of staff and the subsequent costs of recruitment and training of replacement personnel.

Due to the rapid increase in the organisation's workload, certain indi-viduals had also been promoted to positions that they didn't necessarily enjoy, as the following example illustrates:

. . . It was supposed to be a promotion but I don't think it actually was as far as the money is concerned. I got a car, but I never bothered about cars. Before I was a general foreman I always used to have a van. It took me where I wanted to go, I have never been a car lover. I never asked for the promotion, they approached me. I sat on it for at least ten weeks. Everybody was saying you want to do it. My wife was saying you want to do it. So I took it. After three months, I went to see [the director] and said 'this isn't for me, I will have my little van back and I'll give you your car back.

These difficulties in managing the career structures appeared to be symptoms of the short-term outlook towards the resourcing process.

Promising practice career development: total rewards and career route map

Within their very strong hierarchical structure, company D operated an innovative approach to career management. This included a 'total rewards package' and a network of career paths that were guided by job descriptions (see Recruitment and Selection). The total rewards package included competitive salary and benefits, and training and development options. The network of career paths provided transparent progression opportunities. The company Intranet had a career 'route map', which clearly showed the options available from each post. The job descriptions, as alluded to the aforementioned, outlined the minimum and outstanding requirements for each role. These were used to aid discussions on aspired and realistic future moves. They formed a practical tool for benchmarking performance against desired criteria and identifying training and development needs. The system was also used to highlight and fast track those with potential. The success of the Intranet 'route map' was clear; the careers section was the most popular site within the whole of the company intranet.

Performance management and appraisal

The performance appraisal system formed the only formal means of selecting people for promotion. The annual appraisal interview provided an opportunity for discussing potential progression solutions and aided assessing individuals' current job performance, developing personal development plans and recording employees' aspirations and preferences (see Figure 4.14 for the performance appraisal process chart). However, the data gathered

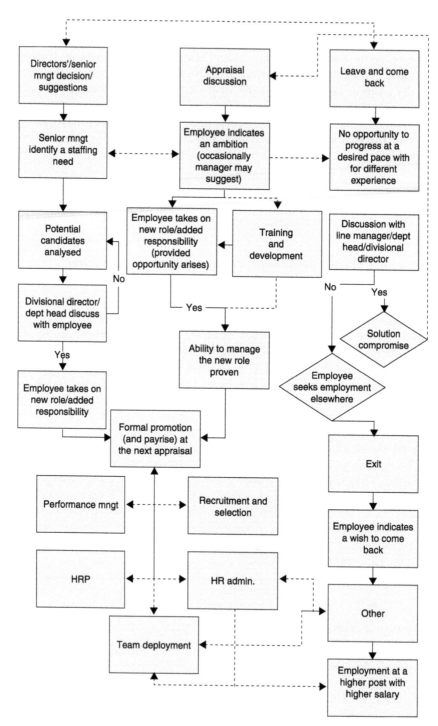

Figure 4.13 Career development (primary case study).

appeared to have little influence over the resourcing decision-making processes:

> The objective part is: am I doing what I am supposed to be doing, if not what am I going to do about it . . . so that is measurable. Then there is the subjective part, which is what does he think of me, what do I think of him, what do I think he thinks of me . . . And that is it. Then we all sign it and put it away for a year!
>
> (Chief estimator)

The fact that only paper copies of the appraisal records were kept may have contributed to the limited use of the data as this may make it too difficult for managers to utilise it effectively. Furthermore, the appraisal linked to the graduate-training programme for junior staff, which also included a performance evaluation tool, was completely removed from the main performance management system. The resulting mass of paperwork that this created, understandably led to managers being reluctant to draw upon such information when making deployment decisions. A well-organised and easily accessible IT-based system would allow for the data to be integrated into the resourcing processes.

Integrated systems

Integrated appraisals and resource management database

Company C's performance management system was geared towards providing information for their resource management database (see HRP and Team Deployment). The system identified employees' skills, their personal objectives, aspirations and preferences, training and career development needs, experience and qualifications. This was seen as an effective tool for encouraging employee involvement as it integrated the employee needs and preferences highlighted within the appraisal system with the HRP, team deployment and other related employee resourcing activities. For the organisation, the system provided information managers and HR personnel could use to identify high flyers and potential succession planning candidates, individual and organisational competencies and capabilities, and how well their employees shared the company values. Annual appraisals comprised a developmental discussion, which also provided employees with an opportunity to have an informal *job chat* with their managers. This was particularly useful for older employees who had no ambitions for progression but would appreciate a formal thank you for a job well done. After each project, employees also had a project appraisal. This was a discussion on the employees' performance on the project in

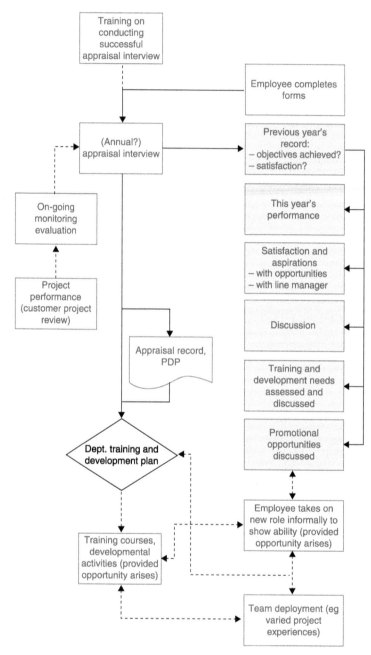

Figure 4.14 The performance appraisal process (primary case study).

question, what he/she would like to do next, what he/she has learned/ not learned, etc. The project appraisal was described as an exceptionally useful mechanism to support team deployment decision-making. It provided information that could be utilised to transfer knowledge across the organisation and supported the employees personal development planning. The developmental aspects of the project appraisal discussion also helped employees and their managers to monitor their progress and the achievement of the targets set in the performance appraisal.

Structured performance management system

Company F's performance management system was highly structured. It was seen as a key HR tool for making sure that employees have the opportunity to speak with their line managers at regular intervals and let them know what they thought of their performance over the past 12 months, as well as what they would like to see happen over the coming 12 months (planning). This also gave the managers an opportunity to tell their employees where they feel that they were at their best, where they feel there might have been room for improvement and how they could improve these areas. HR specialists had the overall responsibility for administering and monitoring the system illustrated in Figure 4.15.

Informal, reactive career development and performance management

An analysis of the performance and career management activities support the view discussed earlier in relation to HRP, recruitment and team deployment; current practice is dominated by informal and reactive processes which undermine the importance of incorporating a comprehensive range of factors to be taken into account in the managerial decision-making. Much of the responsibility for the management and development of an employee's career was devolved to the persons themselves. Managers guided their decision-making only as far as the organisational/project requirements demanded immediate solutions and to which a career move of an identified individual was the optimum solution. Many promotions had been initiated as a response to staffing needs arising from the rapid organisational growth. The side effects and long-term implications of such a response had been overlooked, a symptom typical of the short-term outlook towards the resourcing process. This supports Thite's (2001) outline of the recent developments in careers. However, it contrasts with Baruch's (2003) suggestion that careers are still, to an extent, a 'property' of the organisation, and hence should be managed by them. The normative model of organisational career

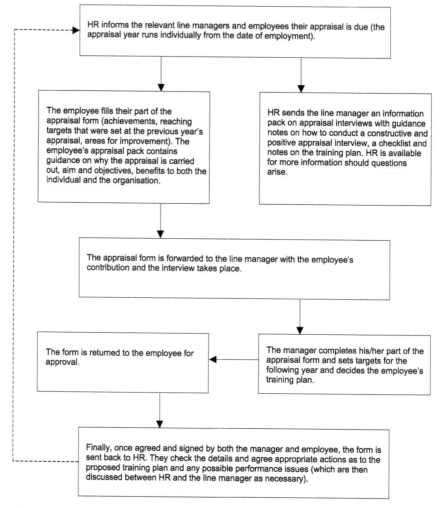

Figure 4.15 A structured performance management system (company F).

management suggests a need for balancing the current somewhat heavy employee responsibility for career planning and providing longer term approach for managing the career structures.

The performance appraisal also enforced the reactive nature of employee resourcing. The system, being entirely paper based, made the utilisation of the collected data very difficult. The duplication of simple procedures understandably frustrated managers, and the overwhelming bureaucracy led to the under-utilisation of the valuable procedures and information. Again, a HRIS could usefully integrate the complementary procedures within a single system and introduce minimal administration requirements via data sharing facilities (Broderick and Boudreau, 1992; Tansley *et al.*, 2001). In summary, a HRIS supported employee resourcing framework that could provide structure for the current informal and reactive performance/career management processes (Tansley *et al.*, 2001) which undermine the importance of incorporating a comprehensive range of factors in the managerial decision-making (see Table 4.11).

HR administration

Much of the HR information was recorded and stored in manual paper files, with the remainder being stored within the computerised systems operated in isolation from each other. For example, a bespoke database had been developed to hold employee records at a company-wide level, but training and development data was held on a separate system administered at a divisional level. Payroll details were held on another system held by the personnel department. This resulted in several members of HR and operational staff having to be involved in any strategic HRM-related decision and the information being transferred across multiple systems. It also made it very difficult to factor in other information such as appraisal data when making resourcing decisions. When discussing the possibilities of introducing an integrated employee self-service HRIS, several managers felt this would prove useful in reducing their administrative workload. Others

Table 4.11 Current performance/career management practice, importance of the function and need for improvement

Current practice	Importance of the function	Need for improvement
• Informal and reactive • Importance of incorporating comprehensive range of factors into the decision-making undermined • Responsibility for career management on employee, managerial guidance only in relation to immediate organisational/project requirements • Long-term consequences overlooked • Appraisal paper-based, process duplication	• Integrated comprehensive strategy for maximising individual, team and organisational performance while facilitating employee career development • Tool for managing the balance between organisational, project and individual employee priorities, needs and preferences • Promotes investment in the development of people • Alignment of the strategic HRM practices with the organisational strategic decision-making • Encourages innovation	• Introduction of structure and longer term planning mechanisms • Introduction of HRIS technology • Recognition for the importance of incorporating comprehensive range of factors into the decision-making • Balance for employee–manager responsibility for career management • Process integration

suggested that it would be useful in promoting movement across divisions when demands on the business required it.

The primary case organisation's HR administration was managed via a complex mix of multiple paper-based and computerised data recording mechanisms. Several members of the HR and operational staff were involved in the collection and transfer of information within as well as between the systems. Thus, the idea of a HRIS as a tool for reducing the managers' administrative workload and encouraging information and knowledge sharing was warmly welcomed. This lends some support for Tansley *et al.*'s conclusion that a HRISs potentially provide the stimulus to effect the required change in employee management practices. However, as the discussion on the systems hereafter shows, parallel to this, a significant cultural shift is required to align attitudes and management practices towards strategic HRM. Without such development, a HRIS is merely 'an electronic filing cabinet'.

Human resource information systems

There are many 'off-the-shelf' human resource information systems (HRIS) solutions available for purchase that may suit the needs of construction organisations. Some systems provide integrated HR/payroll applications where other systems are targeted at specific HRM tasks or functions, such as training and development, performance management or time and attendance management. Certain systems also specialise in web-enabled functionality.

A survey of 100 leading UK construction organisations in 2001 explored the use of HRISs via a questionnaire that asked about the companies' use of information technology applications for HRM-related functions; which HRIS application, if any, they used; the length of time the system had been in place; the functions for which the HRIS was used; and how satisfied they were with the system (Raidén *et al.*, 2001). The results suggested that the use of computers for HR information in large construction companies is broadly similar to the national average. Spreadsheet applications (91.1%), e-mail/internet facilities (77.8%) and database software (68.9%) were most commonly used packages. Just over half (60.0%) of the respondents specified that they used HRIS. Most commonly used HRISs were in-house developed systems (26.7%), with specific commercial applications each being used by less than 10% of the respondents. Considering that only a little over half (60.0%) of the respondents specified using a HRIS, the results demonstrate the use of a wide variety of different systems within the industry.

The same survey was repeated in 2007 to track changes in the use of computers for HR-related functions in construction organisations (Raidén *et al.*, 2008). Although a smaller response rate was received, the results of the second survey revealed that all respondents used word processing and spreadsheet applications to support their HR functions, and e-mail/internet

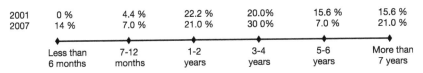

| 2001 | 0 % | 4.4 % | 22.2 % | 20.0% | 15.6 % | 15.6 % |
| 2007 | 14 % | 7.0 % | 21.0 % | 30 0% | 7.0 % | 21.0 % |

| Less than 6 months | 7-12 months | 1-2 years | 3-4 years | 5-6 years | More than 7 years |

Figure 4.16 Length in time the survey respondents had had their HRIS in place.

facilities were used by all but one organisation (94.1%). Majority of the respondents (82.4%) also indicated using a HRIS. This shows an increase in comparison to 2001 data: the percentage of HRIS users in the respondent group then was at 60%. In-house developed systems were still the most commonly used HRISs in 2007. Thirty-five percent of organisations had developed bespoke systems in-house. Also, the use of commercially available systems was similar in range to that recorded in 2001. Thus, the results demonstrate an on-going use of a wide variety of different systems within the industry.

In terms of the length of time, HRIS had been in place, in 2001 most had used the system between one and seven years, as the timeline in Figure 4.16 illustrates. In 2007, the survey recorded more new users, which suggests a marginal increase in the adoption of HRIS software in the construction industry.

The survey also asked about the main uses of HRISs. As Figure 4.17 demonstrates, both in 2001 and 2007 employee records, reports and training administration were most commonly cited functions. While there are slight variations in emphasis (e.g. in attendance, annual leave and equal opportunities monitoring) use of the system is broadly similar. The trend towards increase in the use of systems (also noted earlier in terms of the length in time the respondents had had their system in place) is evident here too.

Finally, the respondents were asked to justify their satisfaction with the system in place. In 2001, 65% of the respondents were satisfied with their HRIS. This represents a mean of 2.98 on a scale from 1 (most dissatisfied) to 5 (most satisfied). In 2007 the figure was higher at 3.57. Figure 4.18 shows the respondents' satisfaction with regard to HRISs serving different HRM functions (mean satisfaction on the 1–5 scale).

Follow-up telephone interviews (in 2001) confirmed that rather than facilitate decision-making through making suggestions as to appropriate deployment, training and career development activities, HRISs were merely used to extract information on employees as required by managers. Informal developmental activities, such as work-based learning, were not recorded in the HRIS, and access tended to be restricted to those with the responsibility of maintaining the data within the systems, rather than the line management staff who could have utilised the information in their day-to-day HRM decision-making.

In 2007, despite the positive responses on general satisfaction and

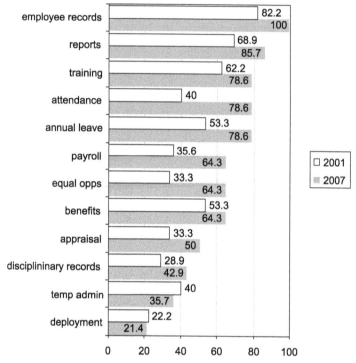

Figure 4.17 Main uses of HRISs 2001/2007.

respondents' satisfaction with HRIS support for specific HRM functions, the additional comments provided by the survey respondents expressed a notable trend in general dissatisfaction with the implementation of HR systems, even where a bespoke package had been developed for the organisation. Some mentioned that the full potential of the systems was not being realised by their company. Few were in the process of changing systems. One company highlighted that they were (still) in the process of introducing new elements of the system into the organisation (phased implementation). This was said to cause frustration, within the HR team and management, as everyone involved wanted the process to be quicker. The complexities of simultaneous implementation/maintenance of the software slowed down the operation. In one organisation, the HRIS implementation process had also identified shortfalls in the existing HRM systems and procedures. These had to be resolved before the completion of the HRIS project. This reflects the observation made earlier in terms of the limits to the potential HRIS held alone in influencing strategic HRM style decision-making. Another organisation had very specific hopes for the HRIS. They were looking into procuring a new system with a view of processing all CITB (Construction-Skills) grant paperwork more easily. High hopes for improved attendance

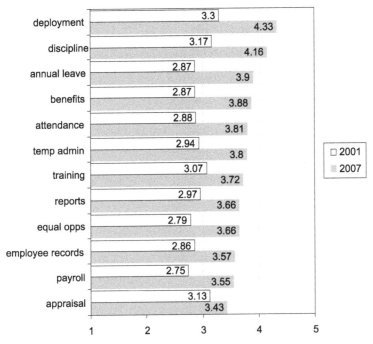

Figure 4.18 Satisfaction in relation to HR functions the system is used for.

monitoring were also recorded: with so many remote workers/sites it is difficult to obtain accurate attendance figures.

Comparison with national trends in HRIS utilisation

In 2001, Raidén *et al.* reported the number of construction companies using HRIS technology to closely correspond with the national average (see CIPD, 2000) with in-house developed systems the most frequently used applications. However, the range of activities that HRISs used in construction was fairly limited and suggested a marked under-utilisation of their capabilities. The industry's application of HRISs for maintaining employee records, managing temporary/fixed-term staff administration and monitoring attendance was 10–20% lower than the national average. Moreover, their application to the monitoring of workforce diversity and equal opportunities was nearly 50% lower than similar sized companies from other sectors.

One of the fundamental arguments for the utilisation of HRIS technology is its ability to support HR managers' decision-making in training provision, employee development and project deployment. Traditionally, these activities have been based on manager's subjective value judgements based on their assessments of human resource capabilities and organisational or project requirements. Despite the obvious advantages of supporting these decisions with an objective needs-based analysis, construction organisations did not appear to exploit HRIS technology to facilitate this process. Despite the sophisticated nature of HRIS software, such packages were used as little more than HR databases, even when developed as bespoke packages in accordance with the operational needs of the organisation.

Seven years on, Raidén *et al.* (2008) note a significant increase in the main uses of HRISs and levels of satisfaction the respondents recorded. A third of the organisations still rely on bespoke in-house systems. Most of the commercially available systems identified were single HRISs which cover several HR functions that are integrated within the system itself and with other IT systems within the wider organisation (see Section 3.3). This is the 'type 2' HRIS in the CIPD (2004) classification, which 21% of their survey respondents use. Therefore, it is very encouraging to note that construction organisations appear to invest in more sophisticated HR software than is the trend nationally across all organisations.

One of the organisations in the 2007 questionnaire indicated using multiple systems with two or more stand-alone HRIS packages that cover different HR functions, but are not integrated with each other or other organisational IT systems. This corresponds with 'type 3' HRIS in the CIPD (2004) classification. Nevertheless, while clearly more sophisticated than simple electronic filing cabinets, these types of HRIS are still usually employed to maintain efficiency and control (Fletcher, 2005).

In terms of length in time, construction organisations tend to be close to the overall averages reported in Ngai and Wat (2006) as shown in Figure 4.19. Although data in Ngai and Wat (2006) is not UK specific, the analysis confirms that construction organisations are working to similar timescales with other industries. This should place them at a competitive level in supporting effective HRM practice.

The range of activities that HRISs use in construction organisations remains restricted to administrative HR functions. In 2001, the main uses were employee records, reports, training administration, annual leave monitoring and benefits administration. By 2007, Raidén *et al.* (2008) noted a particularly significant increase is in three key administration areas: attendance monitoring, administration of annual leave and equal opportunities

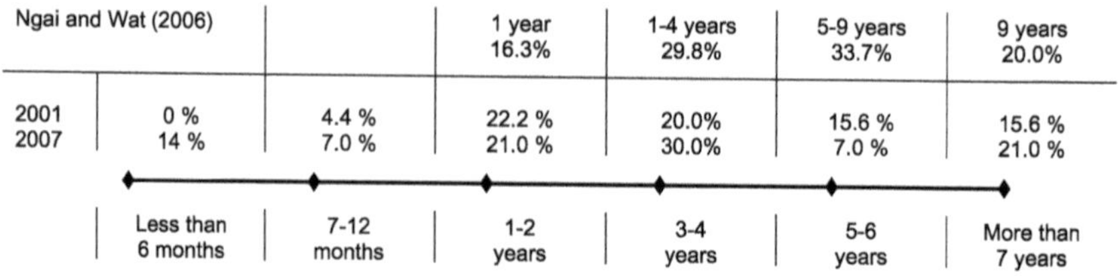

Ngai and Wat (2006)			1 year 16.3%	1-4 years 29.8%	5-9 years 33.7%	9 years 20.0%
2001	0 %	4.4 %	22.2 %	20.0%	15.6 %	15.6 %
2007	14 %	7.0 %	21.0 %	30.0%	7.0 %	21.0 %
	Less than 6 months	7-12 months	1-2 years	3-4 years	5-6 years	More than 7 years

Figure 4.19 Comparison of the length in time the respondents had had their system in place with data from Ngai and Wat (2006).

monitoring (see Table 4.12). These categories of compliance-tracking and process-assurance functionality are of great importance in larger firms, where there is a need to draw from disparate data which are possibly held in separate systems (such as the type 1 single HRIS that covers several HR functions which are integrated within the system itself but not with any other system within the wider organisation). However, taking into account the type of systems in place (in-house developed bespoke packages, type 2 single HRISs which cover several HR functions that are integrated within the system itself and with other IT systems within the wider organisation, and type 3 multiple systems with two or more stand-alone HRIS packages that cover different HR functions, but are not integrated with each other or other organisational IT systems), a marked under-utilisation of their capabilities is suggested.

Table 4.12 outlines further key data from three surveys including the two construction-based questionnaires by Raidén *et al.* (2001 and 2008) and the latest CIPD survey (2005).

These figures also show an increase in the use of HRIS for performance management, including appraisal and attendance monitoring. Figures in 2007 (Raidén *et al.*, 2008) correspond more closely with the national averages recorded by the CIPD (2005). The other area of significant increase noted is equal opportunities monitoring, which also brings the figures for construction organisations more in line with national averages. In 2001, this was particularly low, nearly 50% lower than similar sized organisation in other sectors. This is an important development within a sector that is traditionally occupied by white men, but now increasingly interested in diversifying and thus drawing in a much wider pool of personnel. Effective monitoring of access, opportunities and success of women, ethnic minorities and other minority groups is crucial for ensuring equitable working

Table 4.12 A comparison of the use of HRISs in construction and nationally

		Raidén et al. (2001)	Raidén et al. (2008)	CIPD (2005)
Performance management	Appraisal	33	50	47
	Attendance	40	79	85
Staffing	Human resource planning	–	–	29
	Deployment	22	21	–
	Recruitment	–	–	51
	Temporary/fixed term	40	36	–
Training and development		62	79	75
Equal opportunities/ diversity		33	64	57

practices. Availability of such data is also useful support in the development and evaluation of people management initiatives.

Although the data on staffing activities in the construction industry do not match exactly with the categories in the CIPD survey, close examination of the figures in Table 4.12 reveals noticeable trends. Both, HRP and deployment are strategic HR functions, while recruitment and administration of temporary/fixed-term contracts staff tend to refer to more operational activities. Such grouping indicates that less than a third of organisations nationally and one-fifth of organisations in construction use HRIS to support strategic decision-making. Many more use the systems to assist in the operational tasks, such as recruitment (half of organisations nationally) or administration of temporary/fixed-term contracts staff (a third of organisations in construction).

In summary, cross comparison of the national data with figures from construction industry indicate broadly similar trends across the key function in HRM: staffing, training and performance management.

Explaining the under-utilisation of HRISs in construction

With regard to explanations for the under-utilisation of HRISs in construction organisations (or nationally), little change is evident since 2001. Raidén *et al.* (2001) discussed both general and construction-specific issues that may contribute towards this situation (such as the skills and confidence of HR professionals, and the construction industry's inability to quickly adopt new information technologies). Raidén *et al.* (2008) noted interesting findings in the recent CIPD (2004, 2005) surveys:

> there appears to be a strong correlation between an organisation with good project management skills and knowledge and high satisfaction with a HRIS implementation. Taking that much of the core business in the construction industry is indeed project management, this raises questions about the transfer of knowledge and experience within the organisations internally.

User satisfaction of HRIS software in construction

Despite the narrow utilisation of the capabilities of the HRISs, most respondents seemed generally satisfied with the system in place with clear increase in the overall satisfaction (from 2.98 in 2001 to 3.57 in 2007). However, there were significant variations in satisfaction depending upon the specific HRM-related functions to which the systems were applied. The complex, strategic activities such as deployment attracted very high satisfaction ratings (4.33 in 2007), whereas the more systematic, administrative functions such as payroll and employee records ranked lower (at 3.55 and 3.57, respectively, both 2007). Thus, it appears that the more 'advanced' or strategically focused the activity, the higher the user satisfaction rating. Raidén *et al.* inferred that construction companies that utilise HRISs for complex HRM tasks derive considerable benefit from their application. Such findings contrast markedly with the CIPD's and IES's research, which highlighted significant dissatis-faction among its respondents with the more sophisticated features of HRISs (CIPD, 2000: 5). This could suggest that the complex and dynamic resourcing environment that the construction industry presents is better suited to the application of IT-based systems, as it is precisely in this environment that the maximum benefit can be derived, although such a hypothesis would require investigations.

The surveys did not explore in-depth who had responsibility for managing aspects of the HRM function within the large companies surveyed. How-ever, the results of the telephone interviews conducted in 2001 suggested that it tends to be only HRM staff who have the access to HRISs within the companies studied. This greatly restricts the potential for the exploitation of HRISs if aspects of the HR function are devolved to line management, as appears to be the case in many construction companies. Line managers and employees updating their own records could save considerable time and further focus on the strategic issues of people management within their organisations if they could use HRIS technology effectively. The latest sys-tems have very few limitations to their adaptability and offer the potential to revolutionise the HR function if utilised efficiently.

The secondary case study material added to the understanding of the use of HRISs is through qualitative data. The good practice data given hereafter shows how some organisations attempted to utilise HR technology to achieve the value added, taking the step change from current 'efficiency and control' towards 'enabling insight and partnership' and ultimately 'creating value' (Fletcher, 2005).

Promising practice on HRISs

In-house HRIS

At the time of the research interviews, company A was in the process of developing an in-house HRIS. This was to form a complete HR database for the group and hold all basic personnel data, such as employees' addresses, dates of birth, job titles, absence records and holidays. Inclusion of training and development aspects was planned for the future. Ideally, the system was to include everything the organ-isation need to know about their staff. Despite this, reporting labour/ staff turnover and absence, and ethnic monitoring were seen as the main applications for the system. HR specialists currently manage all strategic HRM-related administration; therefore, it was presumed that they would carry this on but in the future using HRIS. Potentially, such a comprehensive HR database is capable of informing HRP, recruitment and selection, team deployment, learning and development and other strategic HRM-related planning and decision-making.

Human resource planning system

As with company A, company E was in the process of building an internal HRP information system. This was envisaged to include facilities for comprehensive data management and decision-support. The system was to replace the current divisional databases and provide a centralised information source for the group as a whole. It was to run separately from the group's personnel database, but with capabilities for data sharing between the two. The HR department assumed overall responsibility for maintaining the HRP information system as well as the personnel database.

Resource management database

As discussed earlier, company C operated a resource management database. Their performance management system provided all necessary information for this to be run effectively and utilised to support strategic HRM-related decision-making. The HR department held sole responsibility for administering and accessing the system. A HR Officer commented: If we left it to the line managers it wouldn't get done. We need to act as a voice of reason, a neutral party . . .

Change and organisational development

A recent merger had introduced major change within the organisation. Operations of the merging contractors were being integrated and HR-related policies and procedures assimilated. This included adding a smaller division to the operations strategy and deciding a functional focus for the new department. House building was selected with potential for development, although at the time the company had little work in this area. This resulted in temporary decline in the division's workload and consequently to severe staff insecurity. An 18-month delay in agreeing new terms and conditions of employment for the division's employees did not contribute to productive working environment. Communication on progress updates and future plans was minimal. Hence, many of the respondents suggested that integration of the two companies had not been managed effectively.

Relocation of the division's offices was also required. Due to the recent growth of the organisation, room within the region was tight and so the region's training quarters were converted into offices for the new employees. This resulted in the loss of valuable training space and also some resentment from the new employees. In some cases, their travelling distance from home to work had increased by up to 60 miles.

Another cause of concern for some of the employee respondents was the lack of training available. They felt isolated from the region's other divisions, as little training on the organisational policy and procedures had been provided. The estimating staff raised particular concerns with regard to a computerised estimating package in use within the rest of the region. The division's old operating system had been discontinued at the time of the office relation with immediate prospects for introduction of the regional practices. At the time of the research interviews, the estimators had carried out their duties manually for over 12 months, with no near-future predictions for moving onto the computerised estimating package in use within the rest of the region. This was said to have slowed down their work and to have contributed to the poor performance of the division. Consequently, this had contributed towards the reduction in workload, which in turn resulted in some projects becoming overstaffed.

This type of scenario would suggest a reactive approach towards the management of change, with little evidence of higher level strategic planning. This was evident within the organisation's approach towards change management in other areas too. Learning and development appeared reactive rather than proactive in meeting the organisational objectives (Section 4.4) and resourcing decisions/actions were initiated when problems arose, with no forecasting or scenario planning carried out in advance as is evident within the case study hereafter.

Case study: Joint venture

The joint venture project was a high profile £62 million development involving two major UK contractors, the principal case study organisation and another comparable contractor in terms of size, turnover and number of employees. The project duration was scheduled to last three years employing a total of 50 professional staff. At its peak level of production the project required approximately 200 sub-contracted operatives on site.

The overall programme was overseen by the divisional construction director. Similarly to the PFI schools project, matrix management structure applied to the support functions. During the initial design phase and start of the production, the project management and engineering functions were structured according to five principle themes: externals, sub-structures, super-structures, envelope and internals (see Figure 4.20). Each section had an assigned leader with qualities central to the theme (such as design capability/design co-ordination capability/commercial capability/production capability) and a team responsible for their particular element of the building. This structure was initiated in order to avoid 'everyone chasing the same ball', a problem the construction director said to be common in larger projects.

Altogether eight members of the project team were interviewed across a range of professions. This included the construction director, commercial manager, project manager, senior QS, design co-ordinator, site agent and two site engineers.

The initial set up of the project required close teamwork at a senior management level during the planning and project initialisation stages. The start of the construction process followed with intense recruitment and selection, induction and team building phases. Due to the nature of joint venture undertakings, it was an imperative that the production staff worked well together. Not only were the organisations to induct their new recruits into the project organisation that was created to provide singularity and focus for the duration of the project, but also assimilate their existing members of staff. This required extensive team building exercises, with some casualties inevitably occurring. Few existing members of staff were assigned positions they simply did not want to do. Their role had previously involved overseeing considerably smaller contracts independently. Thus, they found it difficult to work as a part of a larger multi-organisation team. Overall, however, the five-theme structure was found successful during the design phase and into the start of the production.

When the project progressed towards the main construction phases the sectional teams begun to divide and power struggles surfaced between them. In response, the team composition was changed to a traditional hierarchical structure with a single leader in charge of the production as a whole. Figure 4.20 shows the change from the five-theme structure to the hierarchical structure.

In addition, approximately half way through the contract, the second partner in the joint venture went into liquidation. The primary case study organisation chose to take the project on as a whole. The lost partner had employed approximately 50% of the project staff. To minimise disturbance to the project progress, the case study organisation decided to offer all of these people temporary contracts of employment for the duration of the project. Despite their efforts, several members of the team were lost to other contractors, who poached key personnel on the project as soon as the news

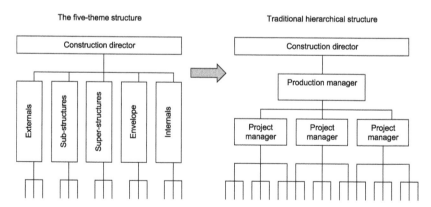

Figure 4.20 The joint venture case study project five-theme design structure and hierarchical production structure.

had been released. Later, permanent employment was offered to almost everyone that had remained with the project through to completion.

Towards the completion of the project a number of employees indicated interests to move onto new projects that the organisation had secured. This was managed to suit the newly acquired projects. The redeployment moves caused a temporary reduction in the joint venture staff and forced the remaining project staff to extend to cover for the lost members' workload. The personnel that had been moved were not replaced, as the project was due to finish shortly. The remaining staff brought the project to completion and were deployed to new projects within the organisation.

In conclusion, the resourcing challenges within the joint venture were as follows:

Team building

- the contract form: joint venture;
- initial set up of the project, which required close teamwork at a senior management level;
- intense recruitment, selection and induction due to the size of the project;
- extensive team building exercises aimed at ensuring the two companies' employees worked well together, the new recruits were inducted into the project organisation, singularity and focus was created for the duration of the project, and the existing members of staff were assimilated;
- staff motivation: ensuring the personnel that had been allocated positions they did not want to do were integrated within the multi-organisation team and performed to required standards.

Change management

- change in the project team structure from the themed structure to traditional hierarchical structure;
- loss of the joint venture partner and subsequent changes in project personnel: temporary contracts of employment offered to all staff previously employed by the lost partner, several members of the team lost to other contractors who poached key personnel on the project as soon as the news had been released;
- changes in project personnel due to employee needs and preferences and the requirements of newly acquired projects;
- temporary reduction in the joint venture staff due to the redeployment moves;
- requirement for the remaining project staff to extend to cover for the lost members' workload.

Misaligned strategy and operational practice

The primary case study organisation's approach to change management in practice contradicted the company's HR policy statement. The lack of communication on the progress of events relating to the merger and severe delay in harmonising the terms and conditions of employment for the new employees resulted in staff dissatisfaction and insecurity. The values of 'open, honest and constructive communication . . .' (see Section 4.2) did not deliver their intended confidence in managerial practice. Significant work–life balance issues also emerged from the increase in travelling distances. Despite the organisation's declaration 'It is important to us to respect our employees' work–home life balance . . .' no initiatives were introduced to help manage this. Furthermore, the lack of systems integration and training on the organisational procedures, especially in relation to the computerised operations systems, had resulted in poor performance of the merging division and in turn temporary overstaffing of projects. Again, this was not in line with the strategic intention communicated by the company's People Statement (Section 4.2), which stated that '[the organisation] undertake to provide each employee with relevant and structured training to provide motivation, job satisfaction and to maximise their contribution to the business'.

This, in line with the given discussion, would further suggest a reactive approach to employee resourcing.

The literature on industry performance improvement agenda and other initiatives that have been adopted to combat the challenges faced by construction organisations [partnering, total quality management (TQM), business process re-engineering, learning organisation and knowledge management] revealed little improvement on the crucial issues of team deployment and other resourcing. Partnering was identified as the only area where success was explicit. The primary case study organisation too had very positive experiences in implementing the principles of partnering. It was incorporated in the corporate strategy . . . to be a market leader in delivering complete construction service through a partnership approach . . .

The principles of partnering were delivered through the informal, 'friendly', organisational culture. Although project work was often stressful, adversarial relations and blaming were discouraged. Senior managers believed that this was their distinct business advantage. Employees were a little unclear of the details the term partnering entailed, but believed the overall approach was deeply embodied within the organisational culture. This created a positive teamwork environment within the organisation and its clients, suppliers and other stakeholders and thus provided a promising foundation for a strategic HRM approach towards employee resourcing. Improving the employee resourcing-specific functions to integrate with the partnering approach would provide a comprehensive methodology for business excellence, in terms of both the internal and external relations.

Total quality management or business process engineering had not been considered by the organisation. Quality was monitored via quality assurance mechanisms. The organisational business and decision-making

processes had evolved from senior management leadership and were operated divisionally within the segregated profit centres. The informality of the
organisational culture influenced much of the success of the set procedures
and the ad hoc operations meant that diversion from the policy/procedure
was not uncommon. For example, when business/project requirements
demanded so, newer employees or freelance staff were left to manage their
projects without procedural guidance or system support. Without the
necessary knowledge of the organisational procedures, their attempts to
record and manage the project information were accepted as appropriate.
A similar approach applied to employee resourcing related processes, as
discussed in detail previously. Cross-referencing this type of practice to
Vakola and Rezgui's (2000) argument on the importance of mapping and
understanding the organisation's current processes, it is clear that the
organisation had not taken the benefits of understanding 'what is going on'
very seriously.

In relation to knowledge management, no explicit initiatives geared
towards managing the organisational knowledge were identified. The localised employee resourcing practices did not take into account the wider
organisational implications and benefits of knowledge sharing, nor did the
training and development practices encourage effective delivery of the learning outcomes to the wider project/divisional community. This undermines
the benefits that effective knowledge management can potentially deliver:
enhanced performance, increased value, competitive advantage and return
on investment. Thus, knowledge management was identified as an area
where substantial improvements could be made via the development of the
employee resourcing framework.

The organisational commitment to training, development and Investors in
People fell closely under the description of the learning organisation (see
Section 4.5). However, this was not recognised as a 'label' for their intended
approach. Perhaps it is for this reason that the concept has received minimal
attention at an applied level. Organisations may not realise their approach
to learning and development has a specific name, and thus find it difficult to
report on the issue.

Employee involvement

Employee involvement in team deployment or other employee resourcing
decision-making was minimal. Indeed, evidence of any kind of employee
involvement practices were limited to financial incentives (in the form of the
bonus scheme and individual pay negotiations) and informal employee–
manager relationships. Within the primary case, organisational culture
encouraged one-to-one contact between line managers and their staff. However, the recent rapid expansion of many of the operating divisions had
raised the importance of effective communications at this level. The contracts managers' workload had increased significantly, which in turn had

also increased the number of employees directly reporting to them. Understandably, managers in charge of up to 150 staff found it difficult to maintain close contact with individual employees. Many employees had noted their manager's increased workload and commented on their unavailability. In addition to the informal employee–manager relationships, many managerial respondents mentioned downward communication. A team briefing structure was said to cascade down throughout the organisation. Divisional directors briefed their senior management teams, who then conducted similar meetings within their respective department heads. The department heads delivered the information to their project managers and senior surveyors and estimators. These personnel then met with their site-based teams. The employees, however, suggested that the team briefings were infrequent and ineffective. This defeated the employee involvement purposes of the team briefing structure, thus making it a pure information delivery mechanism.

This type of employee involvement suggests an approach common throughout all sectors since financial participation, consultation (informal employee–manager relationships) and downward communication are the most common types of employee involvement in use. The primary case organisation had not considered upward problem-solving or task participation, which Marchington (1995) identifies as other powerful techniques. Furthermore, the analysis supports the findings of the construction employee involvement literature, which suggests the prevalence of poor employee involvement practices within the industry (Druker, 2007). This, together with the inconsistency of current employee involvement practice within the organisation clearly suggests a need for considerable improvement in the area if maximum benefits are to be achieved. Accordingly, it is recommended that employee involvement form a key component of a more appropriate employee resourcing framework for the industry. Table 4.13 summarises this.

Summary

This section has discussed the key employee resourcing and related activities that the research findings highlighted central to construction organisations. Although the strategic intention of the primary case study organisation (and in many secondary case organisation too) was found to be positive, this did not translate into effective managerial practice at a project level. Greater employee involvement in and integration within employee resourcing activities could deliver extensive benefits for HRP, recruitment and selection, team deployment, performance and career management, and change management. A HRIS was identified as a potential tool for supporting such development. However, if employee resourcing is to work as an effective facilitator of change via its staffing, performance and HR administration activities, it is an imperative that a broader, long-term strategic view of the

Table 4.13 Current employee involvement, its importance and areas for improvement

Current practice	Importance of the function	Need for improvement
• Not included in the central team deployment function • Limited to financial incentives, informal employee–manager relationships and some downward communication	• Approach to increasing organisational effectiveness through manager and employee collaboration and sharing power and control • Driver for enhanced employee performance and corporate success • Tool for addressing increasing performance demands and mitigating the negative effects of the fragmented project delivery process • Effective way to managing change, ensuring customer satisfaction and encouraging innovation	• More frequent and effective team briefing structure • Introduction of upward problem-solving and task participation • Integration into the team deployment and other employee resourcing functions • Initiating mechanisms for employees to voice their needs and preferences in relation to their project allocation, development, careers and other employee resourcing and strategic HRM-related issues

function is adopted. Its potential as a 'change agent' can only be achieved via integrated, holistic policy and procedures that deliver transparent results and provide a clear focus for the organisation, the projects it manages and people it employs (Taylor, 2005). The next section discusses these requirements of a more appropriate framework in detail.

4.7 The need for integration of the strategic HRM functions

The previous discussion highlights the need for a radical process improvement and integration if benefits from strategic employee resourcing processes are to be achieved. The project case studies contextualised the challenges in team deployment, project allocation, team building, change management, HRP and performance/career management. Areas in particular need of improvement were identified: HRP, recruitment and selection, team deployment, employee involvement and performance/career management. Learning and development was suggested as a crucial component of successful resourcing. Its key role in supporting a culture of learning organisation and effective knowledge management was said to establish a direct link with the resourcing activities. The organisational priorities and project requirements and employee needs and preferences also focused on similar five areas:

1 teams
2 employee involvement
3 careers
4 learning and development
5 planning.

The discussion on HRP, recruitment and selection, team deployment, employee involvement, performance/career management and learning and development indicated that process integration is crucial to strategic HRM approach. An in-depth analysis of the employee resourcing process charts and descriptions however clearly highlighted gaps in the integration of the various aspects of the resourcing function (see Figure 4.21). This illustration technique was developed from a combination of Sparrow and Marchington's (1998) 'integration of HRM systems' model (Chapter 2) and the resourcing objectives and activities after Taylor (2005) (Chapter 3).

Figure 4.21 highlights the current links (black lines) between the different resourcing functions (shaded ovals). Two additional functions, project management and training and development, are included for their vital importance and impact to the resourcing functions. Weaker links that exist but need to be improved (dotted lines) are also represented. Ideally, the figure forms a complete nine-point star diagram. It is evident that many of the links required for this are missing. In fact, only a small minority of the links required for effective strategic and integrated management of the resourcing function are in place. Accordingly, the main requirement for the organisation in terms of process integration is to improve the existing weak links and establish the missing linkages between the different employee resourcing processes. Such developments in the integration of different

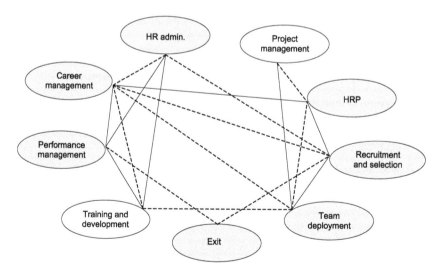

Figure 4.21 Gap analysis of the current HRM activities.

Table 4.14 The performance specification requirements for a more improved employee resourcing framework

	Current practice	Importance of the function	Need for improvement
Human Resource Planning	• Supports organisational strategic intention • Techniques in use: 'what if' scenarios and numerical forecasting • Organisational planning and HRP key themes in resourcing decision-making	• Potential route to organisational flexibility and effective management of change • Can help reduce uncertainty, introduce structure, create order and action and decrease the chaotic nature of a project • HR-business planning integration • Cultural management	• Introduction of HRIS technology • Structured support for the current methods, especially 'what if' scenario planning • Transparent HR-business planning integration • Cultural management support • Effective delivery of the planning outcomes to operational decision-making
Recruitment and selection	• Isolated from HRP • Informal • Fragmented • No vacancy information available group-wide • Word-of-mouth and headhunting prominent methods	• Reconciliation of HRP outcomes with short-term operational conditions • Ensure appropriate supply of skilled staff to the organisation • Contributes to the achievement of business objectives	• Introduction of structured methods • Transparent HRP-recruitment and selection integration
Team deployment	• Ad hoc, reactive to immediate project start requirements or problems • Fragmented, isolated from other resourcing activities • Staff availability number one selection criteria in team formation, employee needs and preferences neglected • Reliance on managers' subjective knowledge • Flexibility managed informally • Team performance measurement weak	• The key to competitive advantage • Most important of all resourcing activities • Powerful motivator • Supports the achievement of the project/organisational goals leading to improved productivity • The intangibles of human interaction separate average performance from outstanding execution	• Longer term planning • Integration with other resourcing activities • Broad range of variables to be incorporated into decision-making • Recognition for employee needs and preferences • Balance between organisational, project and employee factors • Introduction of HRIS • Coherent approach to managing flexibility • Structured team performance and effectiveness measurement framework

(Continued Overleaf)

Table 4.14 Continued

	Current practice	Importance of the function	Need for improvement
Employee involvement	• Not included in the central team deployment function • Limited to financial incentives, informal employee–manager relationships and some downward communication	• Manager and employee collaboration and sharing power and control • Driver for enhanced employee performance and corporate success • Tool for addressing increasing performance demands and mitigating the negative effects of the fragmented project delivery process • Effective way to managing change, ensuring customer satisfaction and encouraging innovation	• More frequent and effective team briefing structure • Introduction of upward problem-solving and task participation • Integration into the team deployment and other employee resourcing functions • Initiating mechanisms for employees to voice their needs and preferences on strategic HRM-related issues
Performance management and career development	• Informal and reactive • Importance of incorporating comprehensive range of factors into the decision-making undermined • Responsibility for career management on employee, managerial guidance only in relation to immediate organisational/project requirements • Long-term consequences overlooked • Appraisal is paper based, process duplication	• Maximising individual, team and organisational performance while facilitating employee career development • Tool for managing the balance between organisational, project and individual employee priorities, needs and preferences • Promotes investment in the development • Alignment of the strategic HRM practices with the strategic decision-making • Encourages innovation	• Introduction of structure and longer term planning mechanisms • Introduction of HRIS technology • Recognition for the importance of incorporating comprehensive range of factors into the decision-making • Balance for employee–manager responsibility for career management • Process integration

| Training and development | • Organisational strategic intention to deliver learning and development
• CPD encouraged
• Training towards professional qualifications and chartered status
• Planning managed via appraisal
• In-house Q-Pulse administration database
• Delivery informal, reactive
• Reactive in meeting the organisational needs
• Unintentional culture of learning organisation (LO)
• Key factor in relation to effective employee resourcing for managers and employees | • Tool for creating sustainable competitive advantage
• Learning and development of organisations and people within them
• Integral element of strategic HRM
• Ensure staff have the skills required for their current roles and can develop those required for future posts
• Motivating/retention factor: training indicates commitment to people and the recipients are more likely to feel valued
• Crucial associate to employee resourcing; key role in supporting LO culture and effective knowledge management (KM) | • Effective delivery of strategic intention to operational practice
• Recognition for employee needs and priorities
• Introduction of HRIS
• Comprehensive and coherent delivery of learning and training activities
• Structure for on-the-job training
• Proactive approach to meeting the organisational needs
• Recognition and active management of the LO culture
• Integration into the employee resourcing activities |

decision-making processes and strategic HRM activities improve information flows through integrated procedures and management practices. Relevant information is gathered and distributed within as well as between functions. Each element is able to draw maximum benefit of its associated components and in turn contribute to the management of other areas via the specialist and/or shared functions. This way, HRP activities are able to distribute overall long-term staffing plans to inform the shorter term team deployment decision-making and recruitment and selection needs (Turner, 2002). Simultaneously, personnel and organisational data gathered via the performance/career management plans and records can be used to support a diverse range of activities, such as internal recruitment and selection, and learning and development (Taylor, 2005). HR administration becomes a single-entry data-processing activity with main focus on data manipulation and distribution (Tansley *et al.*, 2001).

Section 4.5 highlighted four key areas essential for a more improved employee resourcing framework. These were HRP and related recruitment and selection, team deployment, performance/career management and employee involvement. Section 4.4 identified training and development as a crucial associate to successful resourcing through its key role in supporting a culture of learning organisation and effective knowledge management. Due to the dynamic project-based environment, flexibility was also identified as a crucial element. Table 4.14 collates the requirements for a more improved resourcing framework together with the training and development dimension. Flexibility is integrated within the team deployment element.

The integration of these elements requires an extensive policy structure, procedural support and process guidelines, all of which must be augmented with an appropriate organisational culture. The policy structure is required to formally recognise and make public the organisational intent. The procedural support and process guidelines provide practical direction and advice to personnel within the organisation as to the appropriate ways of managing their employee resourcing and other related strategic HRM activities.

There are two practical qualities necessary for the system to be of maximum benefit for managers and staff: easy access and minimal administration. As discussed in Chapter 1, the construction industry presents an exceptionally challenging project-based environment for effective strategic HRM. Pressures for meeting the organisational and project performance objectives are extremely demanding. Thus, a decision-support mechanism that is easy to access and requires minimal daily maintenance is likely to respond to the needs of the organisation better than a complex framework that involves extensive training and upkeep. These must be underpinned by a willingness on the part of the organisation to seek a more systematic and strategic/proactive approach towards managing the resourcing function as part of their broader strategic HRM orientation.

5 A framework for supporting strategic employee resourcing in construction organisations

So far, the book has examined the origins, different components and recent developments in strategic HRM. This has been presented as a useful framework for people management in many industries, including construction. The previous chapter discussed the research findings and results of a study conducted with seven major contractors, and identified a performance specification for a more appropriate approach to employee resourcing decision-making. This chapter introduces the early development of a Strategic Employee Resourcing Framework (SERF), which is intended to support strategic HRM-style employee resourcing decision-making in construction organisations. Clearly, to propose a prescriptive solution to the complex and multifaceted challenges confronting the resourcing function would be both naïve and grossly simplistic. The aim here therefore is summarise some salient issues that must be taken into account when developing resourcing strategies for a large construction company. Firstly, an overview of the framework is followed by details of the design and the main functions. The components are summarised in Figures 5.1 and 5.2. Figure 5.2 also shows the chapter structure. Finally, SERF's potential to support and enhance managerial decision-making is discussed.

5.1 SERF: an integrated HRM model

The research findings and discussion support the conclusions of many studies in strategic HRM: Current employee resourcing decisions tend to focus on meeting the organisational requirements. The importance of integrating employees' needs and preferences into the process is overlooked. The SERF is essentially a conceptual framework designed to illustrate how integrated, technology-assisted employee resourcing decision-making can achieve profitable strategic HRM-style outcomes in complex environments. Table 5.1 summarises the performance specification developed in the previous chapter.

On this basis, SERF has five key elements interconnected via the central team deployment function (see Figure 5.1). This ensures effective process integration, which is crucial to strategic HRM approach to employee resourcing. The other central requirements focus on system flexibility, easy

Table 5.1 SERF performance specification

Key elements	Central requirements	Underlying systems/ support
Team deployment	Process integration	Procedural support
HRP	Flexibility	Process guidelines
Performance/Career management	Easy access	HRIS
Employee involvement	Minimal administration	
Learning and development		

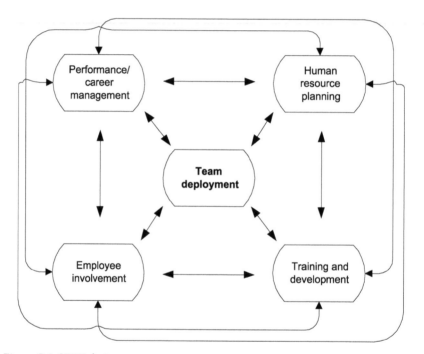

Figure 5.1 SERF design.

access and minimal administration. HRIS was suggested as a tool (underlying system/support) to facilitate these. The HRIS forms the operational/technological structure for the otherwise conceptual model with the purpose of supporting a network of policy, procedural support and process guidelines. This way, horizontal and vertical integration can be achieved.

5.2 Strategic support

In terms of application, the SERF brings together three main elements (see Figure 5.2):

- The user interface (SERF design)
- Decision-making guidelines/protocols
- An operational HRIS tool

These combine to provide a flexible framework, which acts primarily as an analytical tool but also has the potential for supporting decision-making by facilitating effective line management and strategic HRM staff collaboration and information exchange. The comprehensive user interface and policy and procedural guidelines help focus and integrate the organisational strategic HRM and business objectives in the process, communicate them effectively, and also provide structure for managers in their decision-making. The HRIS component allows for effective collection, storage and use of employee data ensuring employees needs and preferences can be easily incorporated into the decision-making process.

Figure 5.2 presents a three-layered conceptualisation of the structure on which basis SERF is organised. The holistic view offers a strategic high-level overview on how the different elements of HRM come together to support effective decision-making. The aim is to communicate the interconnectedness of the functions in an explicit way. Often this level of thinking is abstract to the extent that line managers find it difficult to relate to the overview of the complex network of HRM issues in an organisation. Several examples of this were evident in the data discussed in Chapter 4.

Crucially, isolated HRM activities fail to realise the opportunities offered by process integration. For example, data gathered by performance management systems provide valuable insights into employees' future development aspirations and thus aid planning for training and development. Equally, such data can be used in project allocation and team deployment to match employees to project opportunities.

The mid-layer, which represents the management systems and processes, provides specific support for the five elements of SERF (see Section 5.3). Finally, the operational tool, here fictional ConCo, serves the IT infrastructure that creates seamless data flows and facilitates safe access to information (see Section 5.4).

By carefully integrating the five key elements, the SERF takes into account the diverse and varying needs of different individual employees, as well as those of the organisation and projects. The holistic view provides brief introduction to each element[1], as shown in Figure 5.3.

The team deployment aspect outlines the five processes involved in teamwork. These are creating a balanced team, team building, team performance, project success and re-deploying the team. Each aspect then provides a quick reference to the types of factors that are important to be considered in related decision-making. With regards to creating a balanced team, for example, this includes project requirements, ensuring the team consists of a mix of skills, abilities and personalities, individual employees' needs and preferences, and organisational priorities.

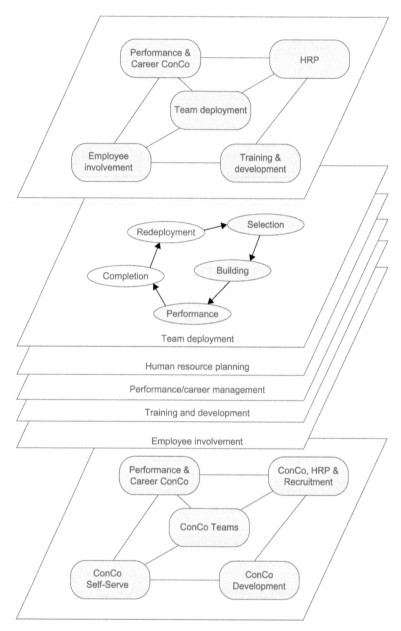

Figure 5.2 The three layer architecture of SERF.

The human resource planning element introduces four crucial stages involved in effective HRP: assessing demand, analysing internal and external supply and integrating the outcomes to the overall business plan. This element includes recruitment and selection as integral aspects of analysing (and

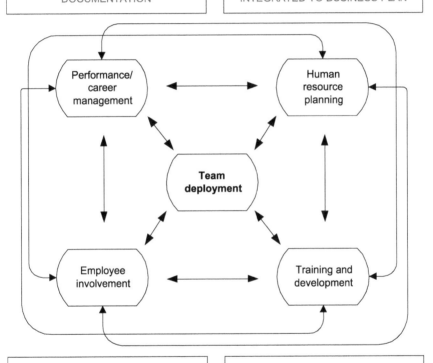

Team deployment
CREATING A BALANCED TEAM
TEAM BUILDING
TEAM PERFORMANCE
PROJECT SUCCESS
REDEPLOYING THE TEAM

Performance career management
CONTINUOUS REVIEW & FEEDBACK
APPRAISAL
INTERVIEW
DOCUMENTATION

Human resource planning
DEMAND
INTERNAL SUPPLY
EXTERNAL SUPPLY
INTEGRATED TO BUSINESS PLAN

Performance/career management

Human resource planning

Team deployment

Employee involvement

Training and development

Employees involvement
INTERNAL NEEDS AND
PREFERENCES INCORPORATED INTO
MANAGERIAL DECISION MAKING,
EMPLOYEE VOICE AND UP-TO-DATE
PERSONNEL RECORDS VIA ConCo
SELF-SERVE
INFORMATION RESOURCE

Training and development
TRAINING PLANS
COURSE DIRECTORY
EXTERNAL OPPORTUNITIES
CPD
GRADUATE DEVELOPMENT
PROFESSIONAL QUALIFICATIONS
LEARNING SUPPORT

Figure 5.3 Overview of SERF.

acting upon) the external supply. The performance management/career development part highlights the importance of continuous review and feedback, and outlines the appraisal procedure. The appraisal overview consists of preparation for the interview together with notes on conducting the interview and related documentation.

The employee involvement component puts emphasis on incorporating the individual employees' needs and preferences into managerial decision-making. It also connects to the HRIS component (of which employee involvement is an integral aspect via the self-service functionality) and points to the information resource provided by the model via the decision-making guidelines/protocols component. Finally, the training and development element includes facilities for supporting employee, team and organisational development.

5.3 Decision-making guidelines and protocols

The decision-making guidelines/protocols in SERF provide structure for managing the five interrelated employee resourcing activities. They state the objectives of each activity/task and outline the main benefits of managing the processes effectively. Clear definitions are included together with summaries of factors that should be taken into account in the decision-making processes. Practical process charts are provided to assist managers in structuring their decision-making, which employees may also find helpful in understanding the rationale behind their managers' decisions.

The following sections present the decision-making guidelines/protocols within the structure of the holistic view. Each of the five key elements are discussed, with the 'schematic SERF' indicating the relevant part of the model.

Human resource planning

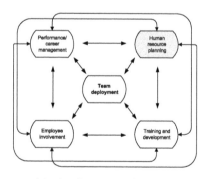

The human resource planning element aims to ensure that the organisational human resources, both in terms of volume of staff and their skills and abilities meet the current and potential business requirements. The main responsibility for the function lies with the organisational senior management, HR staff and divisional managers, although other specialists responsible for business development and marketing may be included in the process. The human resource planning component of the decision-making guidelines/protocols is structured following Table 5.2 outline, which was developed from the best practice/validation case study material and strategic HRM literature.

Table 5.2 SERF human resource planning decision-making protocol outline

Demand	Internal supply	External supply	Integration
• Potential business opportunities • Matching future requirements against existing resources	• Analysis of current staff resources • Analysis of changes in human resources • Analysis of staff turnover • Analysis of effects of changes in the conditions of work	• Analysis of external factors influencing the supply of staff	• Integrating human resource plans into the company business plans

Each column of the table is dealt with in turn including process charts, lists of factors to be considered in the decision-making, details of related processes and the influence in terms of the potential outcomes. Recruitment and selection are also discussed under human resource planning together with the related equal opportunities policy.

Demand

The assessment of the potential business opportunities and matching the future requirements against existing resources focuses on establishing the potential project tender opportunities, identifying gaps between the current staffing levels and those required to fulfil the potential project opportunities, defining the type of recruitment programmes required to achieve the required staffing levels and assessing the feasibility of the potential project opportunities taking into account the economic and social environment. Figure 5.4 lists some factors to consider in assessing the demand for staff.

Internal/external supply

The analysis of the potential internal and external supply of staff include the evaluation of current staff resources, changes in human resources, staff turnover, effects of changes in the conditions of work and the analysis of external factors influencing the supply of staff. Figure 5.5 shows the process.

Internally within the organisation, it is useful to retain records that can be used to show the profile of current staff. Items worth recording may include factors such as the employees' gender, age, education, career development, salary, length of service and personal preferences. Histograms are a useful tool to demonstrate problem areas and changes, for example in the composition of staff complement over time. The charts can highlight the possibilities of growing imbalances, such as the numbers of administrative versus

Qualitative factors	Current skills make-up Current attitudinal measures Current training and development levels Current promotions Knowledge	**Future skills make-up** **Future training and development needs** **Future knowledge requirement** **Succession planning**

Quantitative factors	Current headcount
	  - Where people work (location)
	  - How many work in each location
	Current demographics
	Current business unit headcount
	Current divisional/ departmental/ functional headcount
	Benchmarking comparisons
		Projected headcount (totals)
		Changing geographic patterns
		Forecast demographic changes
		Forecast changes in divisions/departments/ functions
		Future benchmarking targets

	Short-term	Long-term

Figure 5.4 Factors to consider in assessing the demand for staff (Turner, 2002: 98. Reproduced by permission of the publisher: The Chartered Institute of Personnel Development, www.cipd.co.uk).

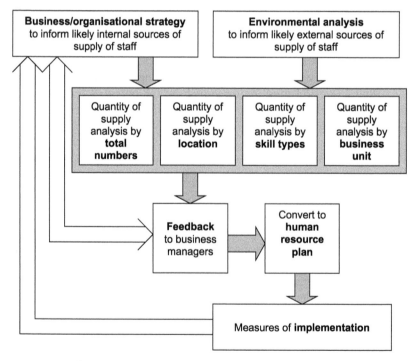

Figure 5.5 Analysis of internal and external supply of staff (Turner, 2002: 109 Reproduced by permission of the publisher: The Chartered Institue of Personnel Development, www.cipd.co.uk).).

management staff, and assist in clarifying staffing plans by focusing on individuals' career paths.

A simple measure of staff turnover for the analysis of the number of new recruits and leavers within the same period of time is

$$\frac{\text{number of leavers in one year}}{\text{average number employed in the same year}} \times 100$$

Exit interviews are an effective way of finding out the main reasons for leaving, and they also aid the collection and maintenance of essential records of starters and leavers.

The analysis of the effects changes in the conditions of work may have assumed that human resource plans are influenced by changes in corporate objectives and environmental changes. Accordingly, this includes the monitoring and evaluation of market and business changes, mapping out alternative opportunities and planning for changes in legislation, such as working hours and the minimum wage.

Succession planning is the primary tool for ensuring the company has the managers available to meet its current and future needs. The power of high-potential, high-performing individual(s) is recognised as a major asset. It forms an important aspect of organisational learning, training and development. The process involves:

• the identification of leadership competencies required to execute organisational strategy;
• the development of organisational structures for identifying emerging leaders;
• creating development paths across functions and business units;
• compliment with formal training and development activities that align with current/forecasted challenges;
• monitoring the engagement, loyalty and critical talent in the organisation.

Analysis of the external factors influencing the supply of staff include factors such as the population density in the area of the potential project, local unemployment levels in the principal professions (and trades), current competition from other companies in the area and likely future competition, availability of short-term housing within the area, the impact of legislation (for example working hours) and specialist and trade contractor arrangements.

Recruitment and selection is the process of attracting and employing the appropriate candidates to fill the gaps identified within the human resource planning process (see Section 3.3). Recruitment focuses on attracting a pool of suitable candidates for the available vacancies. Selection seeks to identify the candidate(s) that best suit the vacancy. The related equal opportunities policies ensure that all candidates within the recruitment and selection processes (and other aspects of strategic HRM) are treated equally and fairly, with decisions being based solely on objective and job related criteria.

Integration

The integration of human resource plans into the company business plans consists of an on-going rigorous process. This includes the continuous cycle through the following:

- Demand forecasting – Estimating staffing needs by reference to the corporate plan.
- Supply forecasting – Estimating supply of staff in the context of current and future supply.
- Forecasting requirements – Analysis of balance between demand and supply so as to be able to predict deficits and surpluses.
- Productivity and cost analysis – Process re-engineering to eliminate wasteful practices where necessary.
- Action planning – Preparation of plans to manage recruitment/set in motion a programme of reduction of human resources.
- Budgeting and control – Setting human resource budgets and monitoring them against the plans.

Team deployment

The team deployment aspect of the SERF operates through a cycle of team selection, team building, team performance, project success and re-deployment, as illustrated in Figure 5.6. It aims to ensure that the teams are balanced in composition, high-performing and effective within which people are motivated and work well

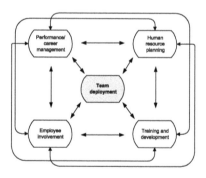

together towards achieving a common goal. The main responsibility for the function lies on the divisional senior management, project managers, site agents/managers, estimators and senior engineers, although any other personnel responsible for project allocation or involved with a project may be included in the process.

Team selection

The team selection focuses on creating balanced teams taking into account the organisational and project and individual employees' needs and requirements. The guidelines outline five approaches to team selection:

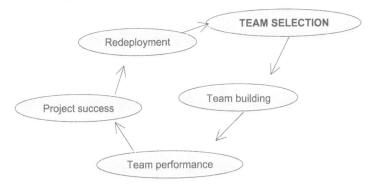

Figure 5.6 SERF team deployment cycle.

interpersonal approach, role definition approach, the values approach, task-based approach and the social identity approach (Table 5.3). Organisations may find particular approaches to suit different circumstances more effectively than others. However in general, the social identity approach is recommended as the most comprehensive method that takes into account the organisational, project and individual employee priorities, requirements, needs and preferences.

Team building and performance

The need to build effective and synergistic teams cannot be overemphasised in construction. High-quality teamwork stimulates innovation and encourages employee commitment, both of which are important competitive levers for construction firms (see Baiden, 2006). *Team building* focuses on establishing synergy between the team members and encourages careful monitoring of the team's integration. Teams often go through a cycle of forming, storming, norming and performing (see Tuckman, 1965). This part also points out some common warning signs of potential sources of conflict and provides useful leader tactics as to how to manage the situation effectively (Table 5.4). This reveals how the team's status is manifested in the ways in which they work together.

Team performance outlines policies and procedures for recognising high achievement and managing performance. Means for recognising high achievement include financial rewards, such as bonuses, pay increases and/or upgrades on car, as well as non-monetary rewards. The non-monetary rewards, such as praise, increased responsibilities leading to promotional opportunities and/or time off work following a successful early completion of a project, may result in longer term satisfaction if operated consistently and on an on-going basis.

Improving poor performance is a challenging task. An individual often under-performs as a result of peer pressure within the team, resistance

Table 5.3 Five approaches to team selection (after Margerison and McCann, 1991; Belbin, 1991, 1993; Hayes, 2002)

Approach	Aim	Main benefits	Downside(s)
Interpersonal approach	High levels of social and personal awareness between team members	Understanding of one another's personalities and ability to communicate helps people work together Team members see each other as 'us' An atmosphere of mutual trust	Lack of recognition of the organisation's needs (strong focus on team)
Role-definition approach	Clarify individual team members role expectations, and the norms and shared responsibilities of the group	Team becomes aware of itself as a working unit and is able to operate effectively and efficiently as members have clear understanding of their role and responsibilities	Lack of recognition of the organisation's needs (strong focus on team)
Team roles/team management wheel	Individual team members play different roles collectively covering all necessary tasks	Successful team consists of a mixture of different individuals	Unable to cope well with individual flexibility
The values approach	Negotiated and shared understanding, values and aims between team members as a group	Team works together efficiently and members are able to see how each individual's activities contribute to the team as a whole	Should all team members not share the vision they are likely to be working at cross-purposes or unable to reconcile conflicts
Task-based approach	Emphasis on the team's and individual team members' tasks (rather than personalities)	Information interchange and realistic analysis of resources, skills and technical aspects of the tasks	Lack of recognition of team members personalities and the interpersonal relationships between them
The social identity approach (a combination of all the aforementioned)	Create a strong sense of unity and belonging Climate of mutual understanding amongst the team members Focus on how and why people can feel proud of belonging to their team	Strong sense of unity will motive the team members to work together and co-operate to achieve their goals The climate of mutual understanding makes everyone aware of the contributions of others, and helps them see how the different parts contribute to the success of the team as a whole Team's contribution to the organisation as a whole visible	

Table 5.4 Common warning signs of potential sources of conflict and useful leader tactics to managing the situation effectively (Holpp, 1999: 84)

Warning signs	*Leader actions*
Forming stage • Little communication • Questioning the purpose of the team • Low trust or commitment • Challenging the leader • Unfocused brainstorming • Disagreement as to problem(s) • Too much talking and wondering • No one takes responsibility for action • Seeks simple/easy solutions • Underestimating problem difficulty	• Select members one by one, with care • Explain purpose of the team • Present clear problem statement • Set goals, timetables, etc. • Maintain sense of urgency • Agree on ground rules for meetings (such as duration, structure, etc.) • Coach problem members outside the meeting • Follow up on specific assignments • Get senior management support/ involvement (where felt necessary/ beneficial)
Storming stage • Cliques begin to form • Unrealistic expectations arise • Members develop at different levels • Realisation of problem difficulty (and 'panic') • Desire to delegate problem upwards • Unwilling to challenge or confront	• Encourage differing points of view • Keep focused on time and goals • Break down larger problems • Seek small successes • Coach members individually • Allow conflict to surface, manage carefully
Norming stage • Arguments occur for no reason • Anger is directed towards team/team leader(s)/management • Team sees the organisation as 'us and them' • Talk is substitute for action • Subgroups go in their own direction – loss of focus • Unanticipated problems break down momentum	• Challenge the group to conduct analysis and resolve disagreement • Move from directive coaching to supportive leadership style • Share leadership duties • Insist members share responsibilities • Use selected tools and techniques religiously • Stick to your goals and time tables
Performing stage • Team takes on too much • Members resist leadership • Members operate autonomously (with little interaction/co-operation with other team members) • Team communication breaks down • Members resist boring/routine work • Team runs out of motivating/ stimulating situations	• Allow the team to set its own course • Enforce regular meeting schedule • Make frequent presentations/check-ups on progress • Make clear 'the bigger picture', demonstrate contribution to the whole (organisation) • Move towards self-managing team approach

to changing work roles, inefficient team communications, differences in working/management styles, coping [or more importantly not coping] with unrealistic expectations, the individual/team wanting to do too much too soon, an over-focus on results (ignoring process) or blaming managers for everything (see Holpp, 1999: 85–99 for an outline of the common characteristics of these problems and recommendations for correcting the behaviour and supporting improved performance).

Project success and team re-deployment

Project success looks at the multiple criteria that are used to determine the success of the team and project and how these can be evaluated throughout the life-cycle of the team/project. Typical project success criteria in the construction industry include time, cost, quality, people and relationships. Their evaluation can be carried out via careful and structured monitoring throughout the project duration, and in completion via project close down meetings and customer project reviews.

The *re-deployment of the project team members* part outlines different strategies for re-deploying the project staff. This includes the re-deployment of the team as a whole to a new project, re-deploying clusters of the team members together to new projects and dismantling the team and re-deploying each individual to a new project. Each have their benefits as well as downsides and suit different situations differing degrees. Accordingly, a method most suitable for achieving the balance between the longer term organisational strategic objectives, immediate project requirements and individual employees' personal and career needs, should be selected.

Performance/career management

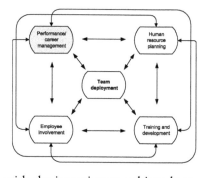

The performance/career management guidelines provide a rationale for effective two-way process highlighting the importance of a strategic and integrated approach. The *strategic* nature of performance/career management is highlighted in that it is concerned with the broad issues facing the business by taking into account the environment within which the organisation operates with the intention to achieve long-term goals. It is *integrated* by means of vertical, functional, human resource and individual integration. The vertical integration links align the business, team and individual objectives. The functional integration links the operational strategies of different parts of the business together. Human resource integration aligns the different aspects of human resource management to achieve a coherent approach to the management and development of people: the strategic HRM approach (Chapters 2 and 3). Finally, the integration of the individual employees' needs incorporates them with those of the organisation/team as far as it is possible. The responsibility for the overall management of the process is placed on the senior management team, operational line managers and HR staff, although each individual employee's role and contribution to effective management of the system is highlighted.

The module includes process charts and tables of factors that are important to be considered in managing performance (see Figure 5.7). Continuous review/feedback together with the appraisal process, personal development plans, job descriptions and job profiles are explained. Career management charts are also included to guide employees in thinking about their longer term development (see Figure 5.8).

Employee involvement

The employee involvement component defines the different types of employee involvement highlighting the benefits of participative management style. It includes policies on work–life balance and flexible working arrangements with details on employee assistance programmes, which offer an independent source of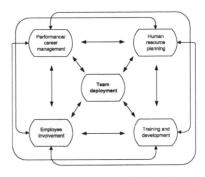
information and advice for those experiencing pressures or difficulties at work or home.

The component is aimed at increasing employees' understanding of the organisation, enabling them to influence decisions, making better use of their talents and encouraging their commitments to the goals of the organisation. Marchington's (1995) five types of employee involvement table (see Section 3.3; Table 5.5) can be used as a basis for developing organisation/division/department specific initiatives.

Practically, the web-enabled SERF is the main tool for supporting employee involvement in many areas. The system includes:

- a facility for employees to input their personal priorities in the team deployment support tool;
- recording and monitoring facilities for training need;
- career management support and advice on promotional opportunities;
- menu for flexible benefits.

The system also holds a comprehensive range of information on company

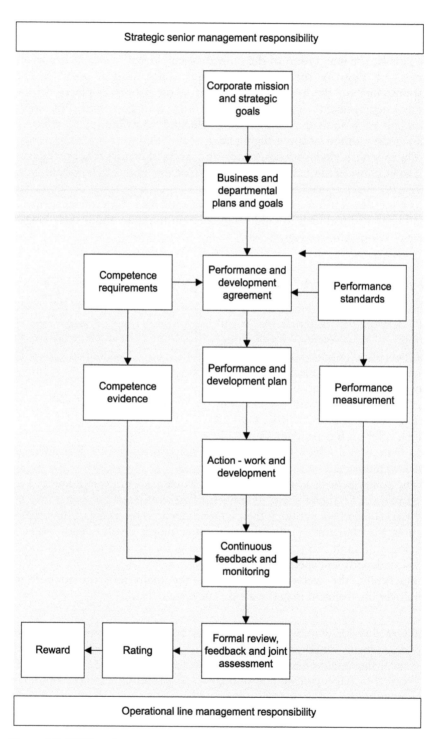

Figure 5.7 SERF performance management process.

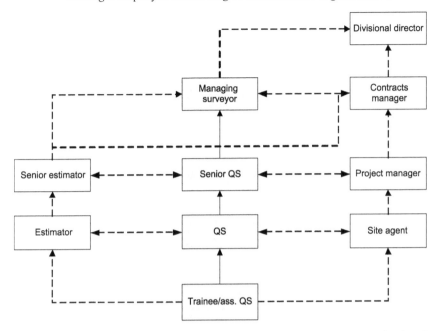

Figure 5.8 SERF career development options for quantity surveying and related routes.

policy and procedures. Employees are advised to refer to this source of information first and then approach appropriate line manager(s) or HR personnel with any questions/suggestions.

Work–life balance is integrated as a key element of the employee involvement component, as it has been found to be an area of increasing concern within the industry (Section 3.3). This is because of the following:

- More people have to juggle responsibilities at home and in the workplace.
- Information and communication technologies have helped to increase employees' performance, but also increased their work intensity.
- More people belong to dual income families where women return to work after having children.
- More men wish to spend time with their families and take on greater responsibility for rearing their children.
- More families are taking on caring responsibilities for elderly relatives as the population ages.
- Employees with no dependants may have commitments within the community, or they may want time to travel, study or engage in leisure activities.

Table 5.5 Five types of employee involvement (Marchington, 1995; Corbridge and Pilbeam, 1998: 332–334)

Type of EI	Objective	Techniques
Downward communication	Managers to provide information to employees in order to develop their understanding of organisational plans and objectives	Formal and informal communications: reports, newspapers, videos, presentations, team briefings
Upward problem-solving	Utilise the knowledge and opinions of employees to, for example, increase the stock of ideas within the organisation, encourage co-operative relationships and legitimise change	Suggestion schemes, total quality management and quality circles, attitude surveys
Task participation	Encourage employees to expand the range of tasks they undertake	Job rotation, job enrichment, teamworking, empowerment, semi-autonomous work groups
Consultation and representative participation	An indirect form of employee involvement, aiming to support effective decision-making, air grievances, 'sound out' employee views on organisational plans	Joint consultation, discussions between managers and employees/ their representatives
Financial participation	Relate the employees' overall pay to the success of the organisation with the assumption that employees will work harder if they receive a personal financial reward from the organisation's success	Profit-sharing schemes, employee share ownership plans

Flexible benefits menu, where employees can select the benefit(s) best suited to their needs from a choice of benefit options, is aimed at meeting the diverse needs and expectations of employees in support of the business. This includes flexible work patterns, such as part-time working, flexi-time, job-sharing, annualised hours, working from home (part-time) and term-time working. Various leave options support this: extended leave, paid paternity leave, career breaks for carers, sabbaticals, study leave and secondments (within career development programme or community support).

Training and development

The training and development component includes information on organisational training plans, external
development opportunities and continuous professional/personal development, graduate development and
professional qualifications together
with training course directories and
learning and development support.

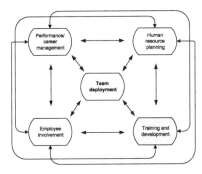

El-Sawad's (1998) 'hard' and 'soft' elements of human resource development (see Section 3.2) are integrated to provide a comprehensive framework
for individual and organisational learning and development. The training
cycle is used to support the planning, delivery and evaluation of the learning
and development outcomes. This helps to ensure that the training needs
analysis is carried out regularly at an organisational, job/occupational and
individual level. As outlined in Section 3.2, comprehensive assessment of the
training needs at multiple levels (organisation/role/individual) ideally leads
to measures of (i) current and future business performance needs and (ii)
current performance and capability levels; and by comparing the two (iii) to
the identification of the current and potential future capability gaps and
(iv) which training and development interventions might effectively address
them within the (v) target population (McClelland, 1993). This helps in
designing and delivering effective solutions, whether traditional packaged
classroom learning experiences are to be offered, computer-based e-learning
encouraged or informal, 'organic', on-the-job learning facilitated via men-
toring/coaching. Finally, effective evaluation of the training and learning
activities demonstrate the impact of human resource development investment
and help modify future planning, design, delivery and evaluation.

Organisational learning is encouraged through individual and depart-
mental/divisional development. The SERF facilitates the recognition and
management of a learning organisation culture, which was previously supported by informal networks. The learning and development modules in
HRISs support this.

5.4 HRIS support

Consideration of the operational HRIS tool assumes interest in practical
application of the model as a managerial decision-support mechanism rather
than a framework for analysis (as a conceptual model). It services the SERF's
underlying database functionality. This activates the conceptual model into
an effective decision-support tool which holds employee records and organisational information readily available for use via the web. Several commercially available applications provide facilities for supporting the model. Here
a worked example, termed 'ConCo', is used to illustrate the potential.

Table 5.6 Benefits of ConCo Central human resource planning, team deployment and performance/career management

Human resource planning	Team deployment	Performance/career management
• Flexible, multi-level framework for analysing current and future skill and competency requirements vs potential supply • Structure for succession planning and gap analysis • Promoting an open, proactive environment	• Building objectivity into the subjective managerial decision-making • Increased employee involvement • Automating processes across the organisation/divisions • Promoting an open, proactive environment	• Automating processes across the organisation/divisions • Creating valuable business intelligence via effective use of employee/ organisational data • Promoting an open, proactive environment

ConCo's employee resourcing support comprises of five components, which serve the five different parts of the employee resourcing function/ SERF (human resource planning, team deployment, performance/career management, employee involvement, training and development). Access to the system is controlled by a strict password protocol, which allows users (managers, administrators *and* employees) to securely view/edit relevant documentation. The integrated 'black box' function ensures confidentiality protecting personal data and details. The five employee resourcing modules are discussed hereafter.

ConCo Central supports the human resource planning, team deployment and performance/career management activities. This is achieved via fields that help administer job profiling and history, competency analysis, employee appraisals and performance assessment, analysis of key performance indicators, workflow, succession and career planning, monitoring and evaluation of business objectives underpinning organisational development and project planning. It is the most comprehensive of the five employee resourcing related modules of ConCo and thus provides a flexible, multi-level framework for analysing the current and future skill and competency requirements vs potential supply, and a structure for HR planning. Table 5.6 summarises the benefits of ConCo Central.

ConCo's recruitment and selection facility helps with vacancy and applicant profiling, applicant search by multiple criteria, monitoring of key vacancy milestones and interview administration. It includes automated address recognition which enables accurate and fast processing of applicant details. The benefits of using information technology to assist in the recruitment and selection of potential candidates can ensure efficient, timely correspondence throughout the entire recruitment process, controlling and managing recruitment budgets and automating processes across the organisation.

At the core of ConCo is the system's database, which holds all the human resources and related information on the company employees. Data are held on employees' current position and work history. Specific absence management and sick pay monitoring facilities link in with pay administration and managing family and special leave entitlements, health and safety records (including risk assessments and accident reports), annual leave, work patterns including flexible working arrangements. Benefit administration (and flexible benefits menu) also link with payroll and employee share options, pensions, company car management, performance management tools and records where they incorporate performance-related pay elements. Such a comprehensive database allows for salary modelling, tracking changes and audit trail as well as effective monitoring of absence and compliance with regulations and data entry. Overall, accuracy is improved with less bureaucracy, processes automated across the organisation and divisions and reserves of organisational knowledge created.

Employee involvement feature further enhances the potential use of this data by providing employees with 'self-service' functionality. Employees and managers are empowered to access, analyse and interact with live information about themselves and their teams. The module is bound to improve data accuracy: responsibility for correctly entering and updating basic details lies with those who have direct interest in the data, the employees themselves. This should improve company-wide communication and generate operational cost savings. The module also provides access to a wide range of company information, including organisational policy and procedures. This helps in the following:

- publishing company-wide static information, such as policies and procedures;
- providing the ability to generate 'live' information, such as vacancies and training schedules, directly from the SERF;
- producing on-screen forms to allow employees and managers to edit live data;
- meeting legal requirements on holding employees' personal data.

There is also a learning and development support tool in ConCo. This includes facilities for course and event administration and evaluation, resource booking, training history up-dates, appraisals, managing training budgets and recording continuous professional development. By effectively managing employee development, the learning and development module can help to improve staff retention and ensure maximum benefit is received from training/development activities by both the organisation and its employees. Since it is accessible via the employee self-service facility, employee engagement is encouraged.

If fully operational, the different components of ConCo can be accessed directly through a web-based interface.

5.5 Summary: the contribution of SERF to effectively managing the strategic HRM challenges in the construction industry

In relation to employment relations, the main benefit is derived from effective provision of consistent information within the organisation, coherent management practice throughout the geographical regions, operating divisions and departments and transparent organisational culture. Human resource development support is provided in the form of clear and transparent information on the varied opportunities available and effective management of the individual and organisational development activities. The model also provides a communication channel, which circumvents the line management structure. This could help to diversify learning activities and promote cross-project and cross-function transfer of knowledge.

In terms of the employee resourcing function, the model facilitates structured and informed decision-making. It ensures that good employee relations are maintained through transparent organisational processes, open communications and extensive employee involvement. The model also supports the employees' career development and management by providing an initial point of contact/information for exploring the opportunities available. It helps managers in offering advice and demonstrating realistic opportunities in career and developmental discussions with their employees.

Flexibility is inherent aspect of the framework. The HRIS component helps balance the organisational, project and individual requirements for flexibility through the structured decision-support and comprehensive information provision. The HRIS facility provides the underlying knowledge base for the framework and encourages employee involvement via the 'self-service' functionality.

Effective information management is the key contribution of the model beyond the employee resourcing specific functions. The model incorporates a web-enabled user interface and an underlying HRIS mechanism, which correlate with the mechanistic approach to knowledge management. These help in capturing and re-using necessary and useful information. Secondly, the model helps in the transfer of knowledge between projects via efficient allocation of human resources and facilitates effective sharing of company and other information which supports the organic, people-centred approach to knowledge management.

It is important to state that the SERF provides as a stimulus for thinking differently about resourcing (i.e. in a more strategic and systematic way about the way in which the resourcing function influences the efficacy of the HRM approach). While it in no way provides a panacea or solution to the problems faced by construction companies grappling with the complex and interrelated issues explored earlier, it does offer a route to thinking about the such issues in a proactive and strategic manner.

Note

1   The examples presented/discussed here have been developed with the needs of the primary (and secondary) case study organisation in mind. Throughout the book, a contingent and contextual approach to strategic HRM is advocated; thus analysis and application of SERF follows that no universally applicable solution can be achieved. Bespoke adjustment would be required to reflect the individual requirements of the company within, in which it is to be applied.

6 Conclusion

The aims of this research monograph were two-fold: to develop a structured and comprehensive understanding of the current employee resourcing practices within large construction organisations and to offer a framework for implementing a strategic approach to HRM. They were achieved via a review of literature in Chapters 2 and 3, critical discussion of organisational practices in Chapter 4 and the development of the strategic employee resourcing framework (SERF) in Chapter 5. This chapter concludes this book summarising the key findings and achievements of the research, together with a discussion of its limitations and suggestions for possible future directions in both research and practice.

One central theme is apparent throughout the discussion on the research findings in Chapter 4: many of the case study organisations operate in a style that is broadly representative of the IR/personnel type approach to managing people. That is, focus on welfare and administration rather than strategic deployment of resources. However, some aspects of strategic HRM are apparent in the general management style and drive on business needs, flexibility and commitment. Furthermore, the devolution of key people management activities to operational line managers was a particular feature. This may have been one contributing factor in one of the major challenges the employees highlighted: diversity of local practices found at divisional/project level. While many organisations deliberately organise themselves in this way to reflect localised needs, employees deployed across divisional boundaries found it very difficult to fit in because of the different ways of working. Often, the organisational strategic intent was very positive with board level directors working as champions for good practice. In the primary case study in particular, the company culture emphasised close working relationships between staff and their managers. Although this was useful in enhancing local commitment, it overlooked the importance of translating organisational strategy and policy into effective managerial practice in projects. Indeed, many line managers viewed the HR function as intrusive and unnecessarily bureaucratic. Again, this may reflect the operational autonomy with which they are used to operating, but it seriously undermines the potential to strategically deploy staff across the business.

Although the context within which construction organisations have to manage the resourcing function is clearly problematic, the challenges are by no means insurmountable. Indeed, the analysis of seven companies' practices has shown that they all use a range of sophisticated approaches in attempting to deal with such complexities. However, managing these processes in an integrated and joined-up way remains a problematic undertaking for construction companies, especially given the wide range of variables identified as needing to be taken into account in the employee resourcing decision-making processes.

The friendly organisational culture and individualistic management style gave many a feeling of confidence in fair practice. Collaborative teamwork and a partnership culture were highlighted as being of paramount importance by both managers and employees in the research sample. Strong leadership and an open and approachable attitude were further priorities for the managers in securing staff retention and achievement of organisational goals. Employees focused on developmental opportunities and work–life balance, although it was clear that each individual's needs and preferences were significantly different to those of their colleagues. The individualisation of the employment contract (and management style more broadly) is one of the key developments in a move away from traditional, paternalistic and personnel management. To align their practices with the strategic HRM paradigm, it is crucial that organisation operates a system which is capable of taking into account the very specific individual needs and preferences. Overall, perhaps naturally, the employee responses tended to centre around operational matters, such as staff induction, promotional opportunities, appraisal and travel, whereas the managers considered a broader range of higher level issues, such as communications, problem solving, commercial aspects of project success and the employee resourcing specific factors.

It is evident throughout the data that the primary case organisation had invested heavily in its learning and development activities and thus successfully created a culture akin to the learning organisation. However, they grappled with some significant issues with regards to their resourcing policy and practice (illustrated by the project case studies and usefully summarised in Table 4.14 in Chapter 4), which fuelled their extensive involvement with the research project and thus created a tangible route to developing strategic HRM in the organisation. As discussed in Chapter 3, strategic HRM comprises three interrelated components: employee relations, learning and development and employee resourcing. True success is achieved in combination of the three.

The study identified many specific issues for the primary case organisation to address in their employee resourcing practice in terms of procedural change, but it also highlighted a need for changes in the organisational culture. Where the informal approach had previously been sufficient, organisational growth now demanded more structured mechanisms. The current approach can no longer form the optimal method when contracts managers each employed 150+ employees directly. In such circumstances one-to-one communication had been reduced significantly. Lack of unified and well communicated organisational policy and procedures had led to extensive confusion, uncertainty and dissatisfaction. Thus, the friendly approach preferred previously had left employees unsupported in new situations with little guidance on ways of managing within the organisational structure. Particularly new employees often need extensive guidance and support managers in change of large numbers of employees are unable to provide. Indeed, while strategic HRM emphasises the devolvement of key operational management activities to line managers, it is here that policy and procedures formulated at strategic level to reflect the organisational culture and then effectively communicated to projects provide support and continuity in approach.

So, the research presented in this book has revealed that although the organisational strategic intention in relation to many companies' approach to people management is positive, current managerial practices did not translate into effective delivery of the strategy, policy or procedures at a project level. The transient and highly competitive environment within which the organisations operated, and production-oriented management style, resulted in informal, reactive and seemingly incoherent employee resourcing decision-making where the organisational and project priorities and requirements dominated the process.

These findings support much of the research on high performance working, which has focused on unravelling the people–performance connection. Like Purcell *et al.* (2003), most of the case study organisations' had clearly focused leadership and 'big vision' for managing people. All were committed to finding ways of inducing and encouraging people to exert discretionary behaviour; 'to go the extra mile'. However, policies need effective implementation and it is in operationalising this strategic intent that the line managers' role is central. This was the single biggest problem identified in the research sample. Implementation of policy held back progressive HRM. In response to these conclusions, the SERF was developed. The key elements to a more strategic employee resourcing decision-making were grouped in five key themes:

1   Human resource planning
2   Team deployment
3   Performance/career management
4   Employee involvement
5   Training and development.

SERF incorporates these five employee resourcing activities within a single integrated framework. This enables the development of systems for proactively *managing* the resourcing function where in the past resourcing has often *reacted* to environmental influences.

SERF addresses many of the characteristics which make construction organisations 'different' and render the applicability of strategic HRM particularly challenging (see Section 1.1). Most importantly, it is project focused. Team deployment is the central concern in the architecture of the model. This facilitates effective management of the fluid structure that challenges useful implementation of many other models of strategic HRM. Project is seen as the 'unit of analysis' or point of implementation where different areas of HRM come together. This enables direct delivery of the 'big vision' addressing the challenges of translating policy into practice and maintaining friendly organisational culture. Secondly, planning is of paramount importance in an environment which is highly susceptible to variations in the demand for services and products. At the same time, planning for employee resourcing is very difficult in an environment where each project requires the deployment of a bespoke team. Although no system can take away the uncertainty, managers and employees in the industry have to work with, by integrating human resource planning as an integral element of all decision-making SERF opens up transparent communications. Where the big picture is clearly visible, both managers and employees are able to see how their input and aspirations fit in. Such 'systems thinking' help alleviate the negative effects the chaotic nature of short-term deployment demands may have. In practical terms, this refers to the on-going collection and analysis of organisational and employee data matching project opportunities to potential staffing scenarios. The sophisticated HRISs that many large organisations in the industry have in place provide a useful tool for such planning. One of the fundamental arguments for the utilisation of HRIS technology is its ability to support manager's decision-making. They are able to hold and process complex sets of data and thus make available information from performance management systems, training plans and project planning schedules. Such process integration is espoused by SERF.

Employee involvement supports the autonomous running of the often disparate project sites with significant and necessary degree of independence. It is also important in engaging the employees with their training and development and career development agenda, where much responsibility relies on the individual. Again, many HRISs, such as the example in ConCo (Chapter 5), have extensive capabilities for employee access. Where employees update their data, managers are able to incorporate changing requirements into their decision-making. At the same time, employees benefit from easy access to wide range of information and can participate in on-going performance and development reviews independently. This, in turn, will support the achievement and development of site management team skills and capabilities needed for organising the highly mobile and itinerant labour force and securing successful project outcomes.

These elements combine to form a considered and well-planned employee resourcing function.

The practical utility of the SERF is demonstrated through the primary case study organisation's commitment and enthusiasm to utilise the outputs of the work (and to purchase specialist software to facilitate the implementation across the organisation). This has already begun to have a significant impact within the organisation by providing a direction for developing their resourcing and other strategic HRM practices. As well as being used by senior managers and HR specialists for managing their resourcing activities, the SERF is being used as the framework for an employment handbook in which the processes behind the deployment activities are transparently explained. As the company has no existing knowledge management systems in place, the SERF structure is also likely to support the development of a culture of knowledge sharing within the organisation.

The six other major construction contractors that participated in the study have also derived benefit from the research outcomes. In particular, one company has begun to use the framework to inform its strategic allocation decisions. The model was seen as having great potential to structure their current strategic HRM decision-making. The user interface and diagrammatic outline of the function were seen as particularly useful aspects of the model. The separation of the key elements of the function (human resource planning, team deployment, performance/career management, employee involvement and training and development) was said to clarify the complex interconnected network of decision-making activities and provide clearer understanding of the interdependent nature of the functions. Managers could focus on specific aspects independently yet be reminded of the relationships between them. Other organisations too have expressed interest in further developing the model as a tool for managing the resourcing process.

However, achievement of such success has significant implications for practice and research within the wider community. While prescribing specific solutions to large populations would be inappropriate, the evidence discussed clearly suggests that there are benefits to considering strategic HRM in modern organisations. Although much of the analysis has focused on the specific issues for the primary case study organisation, an important learning point here is that such an analysis is useful in identifying the weaknesses and strengths for developing organisational practices towards strategic HRM. Prior to the research, the case study organisation's understanding of 'where we are' was limited in terms of the performance of their resourcing policy and practice, and now a detailed insight is available. Combined with their sophisticated approach to learning and development achieving the benefits of strategic HRM is a realistic target. Furthermore, there is a learning point in terms of change management: continuous analysis and evaluation are required to ensure that organisations are operating at their premium and thus achieving the best possible outcomes at all times. The in-depth case is a prime example of how organisational growth demanded a change in the strategic direction that set the organisational culture and hence operational practice too.

A significant challenge for the organisation in planning and implementing change is the tendency of people in the construction industry in particularly, but elsewhere too, to resist change. Their commitment to continuous improvement and learning and development are likely to prove helpful in this. However, in supporting construction organisations to adopt a more strategic HRM in general, this must form one of the key areas for research in future.

As with all research, the findings and results discussed in this book have limitations. The main limitations of this study were time and resource constraints and restrictions in the sample size. The research was conducted within a three-year schedule bounded by the requirements of the project. This inhibited longitudinal study of the changes in the informants' views in relation to changes in their personal/organisational circumstances. The findings and results suggested that both organisational/project priorities and employee personal needs are likely to change over time. Thus, any future study should seek to establish the nature and effect of the changes to the way employee resourcing processes are managed.

Although the conclusions of this research were drawn from an extensive data set, the in-depth analysis and discussion on the current approaches to the employee resourcing function was based on the 50 primary case study interviews. The case study material collected external to the primary case and additional quantitative questionnaire data supported the findings and results. As such, the conclusions are broadly representative of the practices found within large contracting organisations within the industry. However, this study focused solely on professional employees within these organisations. Views of the operative staff and those employed within consulting firms could add interesting insights to this research. Furthermore, as the industry is mainly comprises of small firms, an investigation of the employee resourcing practices within small- or medium-sized organisations would add value to understanding the state of strategic HRM within the industry as a whole.

Methodologically, this research was based on mainly qualitative data. Exploring the issues via quantitative methods within a much larger sample would provide interesting points of comparison to the conclusions derived from the interpretation of the qualitative data. Thus, it is important that research is continued in the area.

Other areas in need of further research include an investigation into organisational sub-cultures and their impact on the implementation of the SERF; examination of the nature and impact of the organisational factors and industry characteristics; examination of the nature and impact of the contextual factors; industry-wide assessment of managerial leadership styles and their impact on the resourcing process, and; testing of the hypothesis on the learning organisation.

The findings revealed that the informal organisational culture and localised sub-cultures have a direct impact on the delivery of the strategic intention to effective managerial practice at divisional/project level. The SERF is suggested to provide a structure for the implementation of coherent management practice organisation-wide. Nevertheless, this does not address the question of how the multiple and diverse sub-cultures potentially influence the implementation of strategic HRM or how the SERF would incorporate the current organisational sub-cultures. Thus, an in-depth investigation into the organisational sub-cultures and their impact on the use of SERF as a benchmark would provide useful information for future development of the model and its practical operation.

Secondly, the range of factors the interview respondents highlighted as important to be taken into account in the resourcing decision-making identified a set of organisational variables and industry characteristics that have a direct impact on the resourcing process. Due to the time and resource constraints and employee resourcing specific focus of this research, these aspects could not be explored in detail. Accordingly, an investigation into the nature and impact of these variables on the resourcing process would provide useful data for further development and use of the SERF. In particular, exploration of possible ways of minimising/managing the impact of the organisational factors and industry characteristics would support this study.

Thirdly, the discussion highlighted a set of contextual factors to the employee resourcing process which have an indirect impact on the resourcing decision-making. An examination of the nature and impact of these factors would again add value to the future development of employee resourcing practices in the industry. In addition, an examination of these factors would enhance the comprehensive understanding of strategic HRM issues within the industry beyond the immediate employee resourcing framework discussed in this research monograph. Together with the findings and results of this study, such data could help develop a more holistic view of the complex network of strategic HRM-related decision-making activities in construction.

Fourthly, the discussion identified a management style which considers the organisational priorities, project requirements and individual employee needs and preferences as crucial to balanced employee resourcing decision-making. However, an assessment of the senior managers in the research sample showed that only a few within the primary case study operated a balanced strategic HRM-type management/leadership style. An industry-wide assessment of the prominent management style would provide interesting comparisons to the findings discussed in this study. Potentially, this could also help identify any differences that exist within and between the industry's leading contractors, consulting firms and smaller companies.

Finally, since this research has clearly highlighted company training and development practices and strategies that align with the concept of learning organisation, without organisational recognition for such terminology, even at HR director level, it is important to test the hypothesis more widely.

References

Abraham, S. E., Karns, L. A., Shaw, K. and Mena, M. A. (2001) Managerial competencies and the managerial performance appraisal process, *Journal of Management Development*, Vol. 20, No. 10, pp. 842–52.

ACAS (2003) *Absence and Labour Turnover*, London: The Advisory, Conciliation and Arbitration Service (ACAS).

Adler, P. (1993) Time and motion regained, *Harvard Business Review*, Jan–Feb, 97–108.

Afifi, S. S. (1991) Factors affecting professional employee retention, *Journal of Management in Engineering*, Vol. 7, No. 2, pp. 187–202.

Anderson, S. D. and Woodhead, R. W. (1987) *Project Manpower Management: Decision-making Processes in Construction Practice*, New York: John Wiley & Sons.

Appelbaum, E. and Batt, R. (1994) *The New American Workplace: Transforming Work Systems in the United States*, Ithaca, NY: ILR Press.

Appelbaum, S. H. and Goransson, L. (1997) Transformational and adaptive learning within the learning organization: a framework for research and application, *The Learning Organization*, Vol. 4, No. 3, pp. 115–28.

Appelbaum, E., Bailey, T., Berg, P. and Kalleberg, A. (2000) *Manufacturing Advantage: Why High-Performance Systems Pay Off*, Ithaca, NY: ILR Press.

Armstrong, M. (2001) *A Handbook of Human Resource Management Practice* (8th ed), Kogan Page, London.

Armstrong, M. and Baron, A. (2002) *Strategic HRM: The Key to Improved Business Performance*, London: CIPD.

Arthur, M. B. and Rousseau, D. M. (1996) The Boundaryless Career: A New Employment Principle for New Organisational Era, in Arthur, M. B. and Rousseau, D. M. eds, *The Boundaryless Career: A New Employment Principle for a New Organisational Era*, Oxford: Oxford University Press, pp. 370–82.

Atkinson, J. (1981) Manpower strategies for flexible firms, *Personnel Management*, Vol. 19, No. 8, pp. 30–35.

Atkinson, J. (1984) Manpower strategies for the flexible organisation, *Personnel Management*, August, pp. 28–31.

Baiden, B. K. (2006) *Framework for the Integration of the Project Delivery Team*, Unpublished PhD thesis, Loughborough University, UK.

Ball, K. (2002) IT plays a crucial part in enabling HR departments to report on human capita, *People Management*, 27th June, p. 61.

Barlow, J. and Jashapara, A. (1998) Organisational learning and inter-firm 'partnering' in the UK construction industry, *The Learning Organisation: An International Journal*, Vol. 5, No. 2, pp. 86–98.

Baruch, Y. (2003) Career systems in transition, a normative model for organisational career practices, *Personnel Review*, Vol. 32, No. 2, pp. 231–51.

Beardwell, I. and Holden, L. (1997) *Human Resource Management, A Contemporary Perspective*, London: Financial Times.

Beardwell, J. and Claydon, T. (2007) *Human Resource Management* (5th ed), Harlow: Prentice Hall Financial Times.

Beer, M., Spector, B., Lawrence, P., Mills, D. and Walton, R. (1984) *Managing Human Assets*, New York: The Free Press.

Beldin, R. M. (1991) *Management Teams: Why They Succeed or Fail*, Oxford: Butterworth-Heinemann.

Belbin, R. M. (1993) *Team Roles at Work*, Oxford: Butterworth-Heinemann.

Belbin, R. M. (2000) *Beyond the Team*, Oxford: Butterworth-Heinemann.

Belout, A. (1998) Effects of human resource management on project effectiveness and success: toward a new conceptual framework, *International Journal of Project Management*, Vol. 16, No. 1, pp. 21–26.

Blake, R. R. and Mouton, J. S. (1985) *The Managerial Grid III*, London: Gulf.

Boxall, P. (1992) Strategic human resource management: beginnings of a new theoretical sophistication? *Human Resource Management Journal*, Vol. 2, No. 3, pp. 60–79.

Boxall, P. (2003) HR strategy and competitive advantage in the service sector, *Human Resource Management Journal*, Vol. 13, No. 3, pp. 5–20.

Boxall, P. and Purcell, J. (2003) *Strategy and Human Resource Management*, London: Palgrave.

Boxall, P. F. and Steeneveld, M. (1999) Human resource strategy and competitive advantage: a longitudinal study of engineering consultants, *Journal of Management Studies*, Vol. 36, No. 4, pp. 443–63.

Boyatzis, R. E. (1982) *The Competent Manager: A Model for Effective Performance*, New York: Wiley.

Bratton. J. and Gold, J. (2003) *Human Resource Management* (3rd ed), Basingstoke: Palgrave Macmillan.

Bresnen, M. J., Goussevskaia, A. and Swan, J. (2004) Embedding new management knowledge in project-based organizations, *Organization Studies*, Vol. 25, No. 9, pp. 1535–55.

Brewster, C. (1998) Flexible Working in Europe; Extent, Growth and the Challenge for HRM, in Sparrow, P. and Marchington, M., eds, *Human Resource Management: The New Agenda*, London: Financial Times, pp. 245–58.

Broderick, R. and Boudreau, J. (1992) Human resource management, information technology and the competitive edge, *Academy of Management Executive*, Vol. 6, No. 2, pp. 7–17.

Brophy, M. and Kiely, T. (2002) 'Competencies: a new sector', *Journal of European Industrial Training*, Vol. 26, no. 2/3/4, 165–76.

Buchanan, D. and Huczynski, A. (2006) *Organizational Behaviour* (6th ed), Harlow: Financial Times/Prentice Hall.

Buchanan, D. A. and Huczynski, A. (1997) *Organisational Behaviour* (3rd ed), Hemel Hempstead: Prentice Hall.

Burden, R. and Proctor, T. (2000) Creating a sustainable competitive advantage through training, *Team Performance Management: An International Journal*, Vol. 6, No. 5/6, pp. 90–96.

Chan, P., Cooper, R. and Tzortzopoulos, P. (2005) Organizational learning: conceptual challenges from a project perspective, *Construction Management and Economics*, Vol. 23, pp. 747–56.

Chan, P., Puybarauc, M. C. and Kaka, A. (2001) An Investigation into the Impacts of Training on Britain's Construction Industry over the Last Twenty Years, *1st International Conference of Postgraduate Research in the Build and Human Environment*, 15–16 March, University of Salford, Salford.

Chapman, R. (1999) The likelihood and impact of changes of key project personnel on the design process, *Construction Management and Economics*, Vol. 17, pp. 99–106.

Charoenngam, C., Ogunlana, S., Ning-Fu, K. and Dey, P. (2004) Re-engineering construction communication in distance management framework, *Business Process Management Journal*, Vol. 10, No. 6, pp. 645–72.

Cherns, A. B. and Bryant, D. T. (1984) Studying the client's role in construction, *Construction Management and Economics*, Vol. 2, pp. 177–84.

Chinowsky, P. and Meredith, J. (2000) Strategic management in construction, *Journal of Construction Engineering and Management*, January/ February, pp. 1–9.

CIPD (2000) Users' views on computerised HR systems, *Chartered Institute of Personnel and Development and Institute for Employment Studies Computers in Personnel 2000 Show Guide*, London: People Management.

CIPD (2001) *Labour Turnover, Survey Report*, London: CIPD.

CIPD (2002) *Recruitment and Retention, Survey Report*, London: CIPD.

CIPD (2003a) *Recruitment and Retention, Survey Report*, London: CIPD.

CIPD (2003b) *Labour Turnover, Survey Report*, London: CIPD.

CIPD (2004) *People and Technology: Is HR getting the best out of IT? Survey Report*, London: CIPD.

CIPD (2005) *People Management and Technology: Progress and Potential, Survey Report*, London, CIPD.

CIPD (2007) *Recruitment, Retention and Turnover, Survey Report*, London: CIPD.

CIPD (2008) Demand for working beyond state pension age set to soar, http://www.cipd.co.uk/pressoffice/_articles/_250108PR.htm?IsSrchRes=1 (accessed 24th May 2008).

CITB (2003) Bridging the gap, http://www.citb.co.uk/research/reports/btg-emid/index-nonflash.html

Cole, G. (2002) *Personnel and Human Resource Management* (5th ed), London: Continuum.

ConstructionSkills (2007) *Blueprint for UK Construction Skills 2007–2011*, Bircham Newton: ConstructionSkills.

ConstructionSkills (2008) *Blueprint for UK Construction Skills 2008 to 2012*, Kings Lynn: Construction Skills.

Cook, S. (1994) The cultural implications of empowerment, *Empowerment in Organizations*, Vol. 2, No. 1, 9–13.

Corbridge, M. and Pilbeam, S. (1998) *Employment Resourcing*, Harlow: Financial Times/Prentice Hall.

Coupar, W. and Stevens, B. (1998) Towards a New Model of Industrial Partnership: Beyond the 'HRM versus Industrial Relations' Argument, in Sparrow, P.

and Marchington, M., eds, *Human Resource Management: The New Agenda*, London: Financial Times Management, pp. 145–59.

Cruise O'Brien, R. (1995) Employee involvement in performance improvement: a consideration of tacit knowledge, commitment and trust, *Employee Relations*, Vol. 17, No. 3, pp. 110–20.

Dainty, A. R. J. and Lingard, H. (2006) Indirect discrimination in construction organizations and the impact on women's careers, *ASCE Journal of Management in Engineering*, Vol. 23, No. 3, pp. 108–18.

Dainty, A. R. J., Neale, R. H. and Bagilhole, B. M. (1999) Women's careers in large construction companies: expectations unfulfilled? *Career Development International*, Vol. 7, No. 4, pp. 353–57.

Dainty, A. R. J., Bagilhole, B. M. and Neale R. H. (2000a) A grounded theory of women's career underachievement in large UK construction companies, *Construction Management and Economics*, Vol. 18, pp. 239–50.

Dainty, A. R. J., Bagilhole, B. M. and Neale, R. H. (2000b) The compatibility of construction companies' human resource development policies with employee career expectations, *Engineering, Construction and Architectural Management*, Vol. 7, No. 2, pp. 169–78.

Dainty, A. R. J., Bryman, A. and Price, A. D. F. (2002) Empowerment within the UK construction sector, *Leadership and Organizational Development Journal*, Vol. 23, No. 6, pp. 333–42.

Dainty, A. R. J., Cheng, M. and Moore, D. R. (2003) Refining performance measures for construction project managers: an empirical evaluation, *Construction Management and Economics*, Vol. 21, pp. 209–18.

Dainty, A. R. J., Ison, S. G. and Root, D. S. (2004a) Bridging the skills gap: a regionally driven strategy for resolving the construction labour market crisis, *Engineering, Construction and Architectural Management*, Vol. 11, No. 4, pp. 275–83.

Dainty, A. R. J., Cheng, M-I. and Moore, D. R. (2004b) A competency-based performance model for construction project managers, *Construction Management and Economics*, Vol. 22, No. 8, pp. 877–88.

Dainty, A. R. J., Raidén, A. B. and Neale, R. H. (2004c) Psychological contract expectations of construction project managers, *Engineering, Construction and Architectural Management*, Vol. 11, No. 1, pp. 33–44.

Dainty, A., Green, S. and Bagilhole, B. (2007) *People and Culture in Construction: A Reader*, Oxon: Taylor & Francis.

De Feis, G. (1987) People: an invaluable resource, *Journal of Management in Engineering*, Vol. 3, No. 2, pp. 155–162.

Devanna, M. A., Tichy, N. M. and Fombrun, C. J. (1984) Matching Model of SHRM, in Fombrun, C. J., Tichy, N. M. and Devanna, M. A., eds, *Strategic Human Resource Management*, New York: John Wiley & Sons.

Dowling, P. J. and Schuler, R. S. (1990) *International Dimensions of HRM*, Boston, MA: PWS-Kent.

Druker, J. (2007) Industrial Relations and the Management of Risk in the Construction Industry, in Dainty, A., Green, S. and Bagilhole, B., eds, *People and Culture in Construction: A Reader*, London: Taylor & Francis, pp. 70–84.

Druker, J. and White, G. (1995) Misunderstood and undervalued? Personnel management in construction, *Human Resource Management Journal*, Vol. 5, No. 3, pp. 77–91.

Druker, J. and White, G. (1996) *Managing People in Construction*, London: IPD.

Druker, J., White, G., Hegewisch, A. and Mayne, L. (1996) Between hard and soft HRM: human resource management in the construction industry, *Construction Management and Economics*, Vol. 14, pp. 405–416.

DTI (2003) *Construction statistics annual*, http://www.dti.gov.uk/construction/stats/constat2003.pdf (accessed 5th January 2004).

Dunlop, J. T. (1958) *Industrial Relations Systems*, Carbondale: Southern Illinois University Press.

Eckford, S., Bates, M. and Whitehead, P. (2001) Evaluating the Efficacy of Project Management Training at the Client Contractor Interface Within a Sector of a large Utility, *1st International Conference of Postgraduate Research in the Built and Human Environment*, 15–16 March, University of Salford, Salford, UK.

Eddy, E., Stone, D. and Stone-Romero, E. (1999) The effects of information management policies on reactions to human resource information systems: an integration of privacy and procedural justice perspectives, *Personnel Psychology*, Vol. 52, No. 2, pp. 335–59.

Edward, A. (1997) All systems go, *People Management*, Vol. 3, No. 12, pp. 43–45.

Elkin, G. (1990) Competency-based human resource development, *Industrial and Commercial training*, Vol. 22, No. 4, pp. 20–25.

El-Sawad, A. (1998) Human Resource Development, in Corbridge, M. and Pilbeam, S., eds, *Employment Resourcing*, Harlow: Financial Times/Prentice Hall, pp. 222–46.

El-Sawad, A. (2002) *Human* Resource Development, in Pilbeam, S. and Corbridge, M., eds, *People Resourcing*, London: Pearson, pp. 284–308.

Emmott, M. and Hutchinson, S. (1998), Employee Flexibility: Threat or Promise, in Sparrow, P. and Marchington, M., eds, *Human Resource Management: The New Agenda*, Pitman, London, pp. 229–43.

Ettorre, B. (1993) So, you have 4,000 jobs to fill . . . *Management Review*, Vol. 82, No. 5, p. 6.

Farnham, D. and Pimlot, J. (1990) *Understanding Industrial Relations* (4th ed), London: Cassell.

Fellows, R., Langford, D., Newcomber, R. and Urry, S. (2002) *Construction Management Management in Practice* (2nd ed), London: Blackwell.

Fitzgerald, L. A. and van Eijnatten, F. M. (1998) Letting go of control: the art of managing in the chaordic enterprise, *International Journal of Business Transformation*, Vol. 1, No. 4, pp. 261–70.

Fletcher, P. A. K. (2005) From Personnel Administration to Business-driven Human Captial Management: The Transformation of the Role of HR in the Digital Age, in H. G. Gueutal and D. L. Stone, eds, *The Brave New World of E-HR*, San Francisco, CA: Jossey-Bass.

Florkowski, G. W. and Olivas-Luján, M. R. (2006) The diffusion of human-resource information-technology innovations in US and non-US firms, *Personnel Review*, Vol. 35, No. 6, pp. 684–10.

Foley, D. A. (1987) Attracting new civil engineers – Adapting to changes in work force, *Journal of Professional Issues in Engineering (ASCE)*, Vol. 113, No. 3, July, pp. 221–8.

Fombrun, C. J., Tichy, N. M. and DeVanna, M. A. (1984) *Strategic Human Resource Management*, New York: John Wiley & Sons.

Ford, D. N., Voyer, J. and Gould Wilkinson, J. M. (2000) Building learning organisations in engineering cultures: case study, *Journal of Management in Engineering*, Issue July/ August, pp. 72–83.

Fordham, P. (2007) Market Forecast May 2007, *Building*, pp. 58–59.

Garavan, T. (1997) The learning organization: a review and evaluation, *The Learning Organization*, Vol. 4, No. 1, pp. 18–29.

Gennard, J. and Judge, G. (2002) *Employee Relations* (3rd ed), London: CIPD.

Geroy, G.D., Wright, P. C. and Anderson, J. (1998) Strategic performance empowerment model, *Empowerment in organisations*, Vol. 6, No. 2, pp. 57–65.

Ghosal, S. and Bartlett, C. (1998) *The Individual Corporation*, London: Heinemann.

Gibb, S. (2001) The state of human resource management: evidence from employees' views of HRM systems and staff, *Employee Relations*, Vol. 23, No. 4, pp. 318–36.

Glad, A. (1994) Save that employee rather than fire, *Journal of Management in Engineering*, Vol. 10, No. 4, pp. 16–18.

Goldberg, S. (2003) Team effectiveness coaching: an innovative approach for supporting teams in complex systems, *Leadership and Management in Engineering*, Vol. 3, No. 1, pp. 15–17.

Gray, R. (2001) Organisational climate and project success, *International Journal of Project Management*, Vol. 19, pp. 103–9.

Green, S. D., Larsen, G. and Chung-Chin, K. (2008) Competitive strategy revisited: contested concepts and dynamic capabilities, *Construction Management and Economics*, Vol. 26, No. 1, pp. 63–78.

Greenlaw, P. and Valonis, W. (1994) Applications of expert systems in human resource management, *Human Resource Planning*, Vol. 17, No. 1, p. 27.

Grugulis, I. (2006) Training and Development, in Redman and Wilkinson, eds, *Contemporary Human Resource Management*, Harlow: Prentice Hall-Financial Times.

Guest, D. (1987) Human resource management and industrial relations, *Journal of Management Studies*, Vol. 24, No. 5, pp. 503–21.

Guest, D. (2001) Human resource management: when research confronts theory, *International Journal of Human Resource Management*, Vol. 12, No. 7, pp. 1092–1106

Guest, D. and Conway, N. (2002) Communication the psychological contract: an employer perspective, *Human Resource Management Journal*, Vol. 12, No. 2, pp. 22–38.

Hancock, M. R., Yap, C. K. and Root, D. S. (1996) Human Resource Development in Large Construction Companies, in Langford, D. A. and Retik, A., eds, *The Organisation and Management of Construction: Shaping Theory and Practice*, Vol. 1, pp. 312–21.

Handy, C. (1980) The changing shape of work, *Organizational Dynamics*, Autumn 1980.

Hayes, N. (2002) *Managing Teams: A Strategy for Success*, London: Thomson Learning.

Hendrickson, A. R. (2003) Human resource information systems: Backbone technology of contemporary human resources, *Journal of Labor Research*, Vol. 24, No. 3, pp. 381–94.

Hendry, C. and Pettigrew, A. M. (1990) Human resource management: an agenda for the 1990s, *International Journal of Human Resource Management*, Vol. 1, No. 1, pp. 17–43.

Hiltrop, J. M. (1996) Managing the changing psychological contract, *Employee Relations*, Vol. 18, No. 1, pp. 36–49.

Holpp, L. (1999) *Managing Teams*, New York: McGraw-Hill.

Honey, P. and Mumford, A. (1982) *Manual of Learning Styles*, London: Peter Honey.

Hosie, P. (1995) Promoting quality in higher education using human resource information systems, *Quality Assurance in Education*, Vol. 3, No. 1, pp. 30–35.

Huang, Z., Olomolaiye, P. O. and Ambrose, B. (1996) Construction Company Manpower Planning, in Thorpe, A., ed., Proceedings of the *12th Annual ARCOM Conference*, Sheffield, Sheffield Hallam University, 11–13 September, Vol. 1, pp. 17–26.

Huber, G. P. (1991) Organisational learning: the contributing processes and the literatures, *Organization Science*, Vol. 2, No. 1, pp. 88–115.

Huczynski, A. A. and Buchanan, D. (2001) *Organisational Behaviour_*(4th ed.), Essex: Pearson.

Huselid, M. A. (1993) The impact of environmental volatility on human resource planning and strategic human resource management, *Human Resource Planning*, Vol. 16, No. 3, pp. 35–51.

Hutchinson, S., Purcell, J. and Kinnie, N. (2000) Evolving high commitment management and the experience of the RAC call centre, *Human Resource Management Journal*, Vol. 10, No. 1, pp. 63–78.

Huxtable, J. and Cheddie, M. (2002) *Strategic staffing plans*, Society for Human Resource Management White Paper, SHRM.

Institution of Civil Engineers (ICE) (1992) Management development in the construction industry, guidelines for the construction professional, London: Thomas Telford.

Ive, G. L. and Gruneberg, S. L. (2000) *The Economics of the Modern Construction Sector*, Basingstoke: Macmillan.

Jackson, P. (1999) *Introduction to Expert Systems* (3rd ed), Harlow: Addison-Wesley.

Jashapara, A. (2003) Cognition, culture and competition: an empirical test of the learning organisation, *The Learning Organization*, Vol. 10, No. 1, pp. 31–50.

Johnson, J. R. (2002) Leading the learning organisation: portrait of four leaders, *Leadership and Organisation Development Journal*, Vol. 23, No. 5, pp. 241–9.

Kagioglou, M., Cooper, R. and Aouad, G. (2001) Performance management in construction: a conceptual framework, *Construction Management and Economics*, Vol. 19, pp. 85–95.

Kang, L., Park, I. and Lee, B. (2001) Optimal schedule planning for multiple, repetitive construction process, *Journal of Construction Engineering and Management*, Vol. 127, No. 5, pp. 382–90.

Kinnie, N. and Arthurs, A. (1996) Personnel specialists' advanced use of information technology – evidence and explanations, *Personnel Review*, Vol. 25, No. 3, pp. 3–19.

Kochan, T. A., Katz, H. and McKersie, R. (1986) *The Transformation of American Industrial Relations*, New York: Basic Books.

Kolb, D. and Fry, R. (1984) Towards an Applied Theory of Experiential Learning, in C. Cooper, ed., *Theories of Group Process*, New York: John Wiley & Sons.

Kolb, D. A. (1984) *Experiential Learning*, Englewood Cliffs: Prentice-Hall.

Kolb, D. A. (1996) Management and the Learning Process, in Starkey, K., ed., *How Organisations Learn*, London: ITP.

Kossek, E., Young, W., Gash, D. and Nichol, V. (1994) Waiting for innovation in the human resources department: Gobot implements a human resource information system, *Human Resource Management*, Vol. 33, No. 1, pp. 135–159.

Kululanga, G. K., McCaffer, R., Price, A. D. F. and Edum-Fotwe, F. (1999) Learning mechanisms employed by construction contractors, *Journal of Construction Management and Engineering*, Issues July/August, pp. 215–23.

Langford, D., Hancock, M., Fellows, R. and Gale, A. (1995) *Human Resource Management in Construction*, Harlow: Longman.

Larraine, G. and Cornelius, N. (2001) Recruitment, Selection and Induction in a Diverse Competitive Environment, in Cornelius, N., ed., *Human Resource Management, A Managerial Perspective*, London: Thomson Learning.

Laufer, A., Woodward, H. and Howell, G. (1999) Managing the decision-making process during project planning, *Journal of Management in Engineering*, Vol. 15, No. 2, pp. 79–84.

Lawler, E. E. and Mohrman, S. A. (2003) HR as a strategic partner: what does it take to make it happen? *Human Resource Planning*, Vol. 26, No. 3, pp. 15–29.

Legge, K. (1989) Human resource management, a critical analysis, in Storey, J., ed., *New Perspectives on Human Resource Management*, London: Routledge.

Leopold, J., Harris, L. and Watson, T. (2005) *The Strategic Managing of Human Resources*, Harlow: Prentice Hall Financial Times.

Lester, S. and Kickul, J. (2001) Psychological contracts in the 21st century: what employees value most and how well organisations are responding to these expectations, *Human Resource Planning*, Vol. 24, No. 1, pp. 10–21.

Lingard, H. (2003) The impact of individual and job characteristics on 'burnout' among civil engineers in Australia and the implications for employee turnover, *Construction Management and Economics*, Vol. 21, No. 1, pp. 69–80.

Lingard, H. and Sublet, A. (2002) The impact of job and organisational demands on marital or relationship satisfaction and conflict among Australian civil engineers, *Construction Management and Economics*, Vol. 20, pp. 507–21.

Long, R. F. (1997) Empowering organisations and their employees, *Engineering Management Journal*, Vol. 7, No. 6, pp. 297–303.

Loosemore, M., Dainty, A. R. J. and Lingard, H. (2003) *Human Resource Management in Construction Projects, Strategic and Operational Approaches*, London: Spon Press.

Love, P. E. D., Heng, L., Irani, Z. and Faniran, O. (2000) Total quality management and the learning organization: a dialogue for change in construction, *Construction Management and Economics*, Vol, 18, pp. 321–31.

Love, P. E. D., Holt, G. D., Shen, L. Y., Li, H. and Irani, Z. (2002) Using systems dynamics to better understand change and rework in construction project management systems, *International Journal of Project Management*, Vol. 20, pp. 425–36.

Mabey, C. and Salaman, G. (1995) *Strategic Human Resource Management*, Oxford: Blackwell Business.

Mabey, C., Salaman, G. and Storey, J. (1998) *Strategic Human Resource Management* (2nd ed), Oxford: Blackwell Business.

Macafee, M. (2007) How to conduct exit interviews, *People Management*, Vol. 13, No. 14, pp. 42–43.

MacDuffie, J. P. (1995) Human resource bundles and manufacturing performance: organizational logic and flexible production systems in the world auto industry, *Industrial and Labor Relations Review*, Vol. 48, No. 2, pp. 197–221.

Maloney, W. (1997) Strategic planning for human resource management in construction, *Journal of Management in Engineering*, May/ June, pp. 49–56.

Marchington, M. (1995) Involvement and Participation, in Storey, J., ed., *Human Resource Management: A Critical Text*, London: Routledge.

Marchington, M. and Grugulis, I. (2000) 'Best practice' human resource management: perfect opportunity or dangerous illusion, *International Journal of Human Resource Management*, Vol. 11, No. 4, pp. 905–25.

Marchington, M. and Wilkinson, A. (2002) *People Management and Development: Human Resource Management at Work* (2nd ed), London: CIPD.

Marchington, M. and Wilkinson, A. (2005) Participation and Involvement, in Bach, S., ed., *Personnel Management in Britain* (4th ed), Oxford: Blackwell.

Marchington, M., Carroll, M. and Boxall, P. (2003) Labour scarcity and the survival of small firms: a resource-based view of the road haulage industry, *Human Resource Management Journal*, Vol. 13, No. 4, pp. 5–22.

Margerison, C. and McCann, D. (1991) *Team Management: Practical Approaches*, London: Mercury.

McClelland, S. B. (1993) Training needs assessment: an 'open-systems' approach, *Journal of European Industrial Training*, Vol. 17, No. 1, 12–17.

Miller, M. (1998) Great expectations: is your HRIS meeting them? *HR Focus*, Vol. 75, No. 4, pp. 1–4.

Moore, D. R. and Dainty, A. R. J. (1999) Integrated project teams' performance in managing unexpected change events, *Team Performance Management, An International Journal*, Vol. 5, No. 7, pp. 212–22.

Mullins, L. J. (1996) *Management and Organisational Behaviour* (4th ed), London: Pitman.

Mullins, L. J. (2002) *Management and Organisational Behaviour* (6th ed), Harlow: Prentice Hall.

Mullins, L. J. (2005) *Management and Organisational Behaviour* (7th ed), Harlow: Prentice Hall Financial Times.

Mumford, A. (1995) The learning organization in review, *Industrial and Commercial Training*, Vol. 27, No. 1, pp. 9–16.

Myers, J. and Kirk, S. (2005) *Managing Processes of Human Resource Development*, in Leopold, J., Harris, L. and Watson, T., eds, *The Strategic Managing of Human Resources*, Harlow: Prentice Hall Financial Times, pp. 351–79.

Nesan, L. J. and Holt, G. (1999) *Empowerment in Construction: The Way Forward for Performance Improvement*, Baldock: Research Studies Press.

Nevis, E. C., DiBella, A. J. and Gould, J. M. (1995) Understanding organisations as learning systems, *Sloan Management Review*, Vol. 36, No. 2, pp. 73–85.

Newell, A. (2001) The Organisation that Learns, in Morton, C., Newell, A. and Sparkes, J, eds, *Leading HR: Delivering Competitive Advantage*, London: CIPD, pp. 95–111.

Ng, T. S., Skitmoore, R. and Sharma, T. (2001) Towards a human resource information system for Australian construction companies, *Engineering, Construction and Architectural Management*, Vol. 8, No. 4, pp. 238–49.

Ngai, E. W. T. and Wat, F. K. T. (2006) Human resource information systems: a review and empirical analysis, *Personnel Review*, Vol. 35, No. 3, pp. 297–314.

Nicolini, D. (2002) In search of 'project chemistry', *Construction Management and Economics*, Vol. 20, pp. 167–77.

Nicolini, D. and Meznar, M. B. (1995) The social construction of organisational learning: conceptual and practical issues in the field, *Human Relations*, Vol. 48, No. 7, pp. 727–46.

Nyhan, B., Cressey, P., Tomassini, M., Kelleher, M. and Poell, R. (2004) European perspectives on the learning organisation, *Journal of European Industrial Training*, Vol. 28, No. 1, pp. 67–92.

O'Connor, D. D. (1990) Trouble in the American workplace: The team player concept strikes out, *ARMA Record Management Quarterly*, Vol. 24, No. 2, pp. 12–15.

Odusami, K. T. (2002) Perceptions of construction professionals concerning important skills of effective project leaders, *Journal of Management in Engineering*, Vol. 18, No. 2, pp. 61–67.

OECD (1986) *Flexibility in the Labour Market: The Current Debate*, Paris: OECD.

OECD (1989) *Labour Market Flexibility: Trends in Enterprises*, Paris: OECD.

Ogunlana, S. and Siddiqui, Z. (1999) Factors Considered in Assigning Project Managers to Projects, in Bowen, P. and Hindle, R., eds, *CIB W55 & W65 Joint Triennial Symposium*, Cape Town, South Africa, 5–10 September.

Ogunlana, S., Siddiqui, Z., Yisa, S. and Olomolaiye, P. (2002) Factors and procedures used in matching project managers to construction projects in Bangkok, *International Journal of Project Management*, Vol. 20, pp. 385–400.

Olomolaiye, P. O., Jayawardane, A. K. W. and Harris, F. C. (1998) *Construction Productivity Management*, Harlow: Addison Wesley Longman.

Pearce, D. (2003) *The Social and Economic Value of Construction: The Construction Industry's Contribution to Sustainable Development*. London: nCRISP, Davis Langdon Consultancy.

Pedler M., Burgoyne, J. G. and Boydell, T. (1991) *The Learning Company: A Strategy for Sustainable Development*, Maidenhead: McGraw-Hill.

Pedler, M., Burgoyne, J. G. and Boydell, T. (1988) *Applying Self-development in Organisations*, Hemel Hempstead: Prentice-Hall.

Phillips, B. T. (2003) A four-level learning organisation benchmark implementation model, *The Learning Organization*, Vol. 10, No. 2, pp. 98–105.

Pilbeam, S. and Corbridge, M. (2002) *People Resourcing* (2nd ed), London: Pearson.

Pollert, A. (1988) The 'flexible firm': a model in search of reality (or a model in search of a practice)? *Warwick Papers in Industrial Relations*, University of Warwick.

Priem, R. L., Butler, J. E. (2001) Is the resource-based theory a useful perspective for strategic management research? *Academy of Management Review*, Vol. 26, No. 1, pp. 22–40.

Purcell, J. (1999) Best practice and best fit: chimera or cul-de-sac, *Human Resource Management Journal*, Vol. 9, No. 3, pp. 26–41.

Purcell, J. and Ahlstrand, B. (1994) *Human Resource Management in the Multi-Divisional Company*, Oxford: Oxford University Press.

Purcell, J., Kinnie, N., Hutchinson, S., Rayton, B. and Swart, J. (2003) *Understanding the People and Performance Link: Unlocking the Black Box*, London: CIPD.

Raidén, A. B. and Dainty, A. R. J. (2006) Human resource development (HRD) in construction organisations: an example of a 'chaordic' learning organisation? *The Learning Organization*, Vol. 13, No. 1, pp. 63–79.

Raidén, A. B., Dainty, A. R. J. and Neale, R. H. (2001) Human Resource Information Systems in Construction: Are their Capabilities Fully Exploited? in Akintoye, A., ed., *17th Annual ARCOM Conference*, University of Salford, Salford, UK, 5–7 September, Vol. 1, pp. 133–42.

Raidén, A. B., Dainty, A. R. J. and Neale, R. H. (2002a) Employee Resourcing: Finding the Balance, in Sun, M., Ghassan, A., Ormerod, M., Ruddock, L., Green, C. and Alexander, K., eds, *2nd International Postgraduate Research Conference in the Built and Human Environment*, University of Salford, Salford, UK, 11–12 April, pp. 249–58.

Raidén, A. B., Dainty, A. R. J. and Neale, R. H. (2002b) Employee Resourcing for a Medium-large UK Contractor, in Uwakweh, B. and Minkarah, I., eds, *10th International Symposium of the CIB W65 commission on Organisation and Management of Construction: Construction Innovation and Global Competitiveness*, University of Cincinnati, Cincinnati, OH, USA, 9–13 September, Vol. 2, pp. 1429–45.

Raidén, A. B., Dainty, A. R. J. and Neale, R. H. (2006) Balancing employee needs, project requirements and organisational priorities in team deployment, *Construction Management and Economics*, Vol. 24, No. 8, pp. 883–95.

Raidén, A. B., Williams, H. and Dainty, A. R. J. (2008) Human Resource Information Systems in Construction: A Review Seven Years On, in *24th Annual ARCOM Conference*, 1–3 September 2008, University of Glamorgan, Cardiff, UK.

Redman, T. and Wilkinson, A. (2006) *Contemporary Human Resource Management* (2nd ed), Harlow: Prentice Hall Financial Times.

Reid, M. A., Barrington, H. and Kenney, J. (1992) *Training Interventions: Managing Employee Development*, London: IPM.

Robinson, S. and Rousseau, D. (1994) Violating the psychological contract, *Journal of Organisational Behaviour*, Vol. 15, No. 3, pp. 245–60.

Rousseau, D. M. (1994) Two ways to change and keep the psychological contract: theory meets practice, *Executive Summary for the International Consortium for Executive Development Research*, Lausanne, Switzerland.

Rousseau, D. M. (1995) *Psychological Contracts in Organizations: Understanding Written and Unwritten Agreements*, London: Sage.

Saaty, T. (1980) *The Analytic Hierarchy Process*, New York: McGraw Hill.

Schaffer, R. J. (1988) Manpower planning – make a "moral contract", *Journal of Management in Engineering*, Vol. 4, No. 1, pp. 56–59.

Schein, E.H. (1996) Career anchors revised: implications for career development in the 21st century, *The Academy of Management Executive*, Vol. 10, No. 4, pp. 80–88.

Schirmer, G. E. (1994) Proactive career management, *Journal of Management in Engineering*, Vol. 10, No. 1, pp. 33–36.

Senge, P. M. (1990) *The Fifth Discipline: The Art and Practice of the Learning Organization*, New York: Doubleday.

Senge, P. M. (1991) *The Fifth Discipline: The Art and Practice of the Learning Organization*, New York: Doubleday.

Senge, P. M., Roberts, C., Ross, T. N., Smith, B. J. and Kleiner, A. (1994) *The Fifth Discipline Fieldbook: Strategies and Tools for Building a Learning Organization*, London: Doubleday/Currency.

Serpell, A. and Maturana, S. (1995) A Decision Support System for Construction Human Resources Management, in Pahl and Werner, eds, *6th International*

Conference Computing in Civil and Building Engineering, Rotterdam, The Netherlands, Vol. 2, pp. 1529–35.

Shah, J. B. and Murphy, J. (1995) Performance appraisals for improved productivity, *Journal of Management in Engineering*, Vol. 11, No. 2, pp. 26–29.

Shi, J. and Halpin, D. (2003) *Implementation of Construction Enterprise Resource Planning (ERP) Systems in the Construction Industry*, http://www.fiu.edu/~sazha002/research/erp-paper.pdf (accessed 10th October 2003).

Sisson, K. (1990) Introducing the human resource management journal, *Human Resource Management Journal*, Vol. 1, No. 1, pp. 1–11.

Sisson, K. (1993) In search of HRM, *British Journal of Industrial Relations*, Vol. 31, No. 2, pp. 201–11.

Sisson, K. (1994) Personnel Management: Paradigms, Practice and Prospects, in K. Sisson, ed., *Personnel Management*, Blackwell, Oxford, 3–50.

Sisson, K. and Storey, J. (2000) *The Realities of Human Resource Management: Managing the Employment Relationship*, Buckingham: The Open University Press.

Smithers, G. and Walker, D. (2000) The effect of the workplace on motivation and demotivation of construction professionals, *Construction Management and Economics*, Vol. 18, pp. 833–41.

Soret, N. (2007) *Dismissal*, CIPD, http://www.cipd.co.uk/subjects/emplaw/dismissal/dismissal.htm?IsSrchRes=1 (accessed 24th May 2008).

Sparrow, P. and Marchington, M. (1998) *Human Resource Management: The New Agenda*, London: Financial Times.

Sparrow, P. R. and Cooper, C. L. (2003) *The Employment Relationship: Key Challenges for HR*, Oxford: Butterworth-Heinemann.

Spatz, D. M. (1999) Leadership in the construction industry, *Practice Periodical on Structural Design and Construction*, Vol. 4, No. 2, pp. 64–68.

Spatz, D. M. (2000) Team-building in construction, *Practice Periodical on Structural Design and Construction*, Vol. 5, No. 3, pp. 93–105.

Stewart, D. (2001) Reinterpreting the learning organisation, *The Learning Organization*, Vol. 8, No. 4, pp. 141–2.

Storey, J. (1992) *Developments in the Management of Human Resources*, Oxford: Blackwell.

Strategic Forum for Construction (2002) *Accelerating Change*, London: Rethinking Construction.

Swenson, D. X. (1997) Requisite conditions for team empowerment, *Empowerment in Organisations*, Vol. 5, No. 1, pp. 16–25.

Tansley, C., Newell, S. and Williams, H. (2001) Effecting HRM-style practices through an integrated human resource information system, *Personnel Review*, Vol. 30, No. 3, pp. 351–70.

Taylor, S. (2002a) *The Employee Retention Handbook*, London: CIPD.

Taylor, S. (2002b) *People Resourcing* (2nd ed), London: CIPD.

Taylor, S. (2005) *People Resourcing* (3rd ed), London: CIPD.

Taylor, S. (2008) *People Resourcing* (4th ed), London: CIPD.

Teece, D. J., Pisano, G. and Shuen, A. (1997) Dynamic capabilities and strategic management, *Strategic Management Journal*, Vol. 18, No. 7, pp. 509–33.

Tener, R. K. (1993) Empowering high-performing people to promote project quality, *Journal of Management in Engineering*, Vol. 9, No. 4, pp. 321–28.

The Housing Forum (2002) *Recruitment, Retention and Respect for People, 20 Ideas for Delivering the 3R's*, London: Rethinking Construction.

Thite, M. (2001) Help us but help yourself: the paradox of contemporary career management, *Career Development International*, Vol. 6, No. 6, pp. 312–17.

Tichy, N. M., Fombrun, C. J. and Devanna, M. A. (1982) Strategic human resource management, *Sloan Management Review*, Vol. 23, No. 2, pp. 47–61.

Torrington, D. and Hall, L. (1991) *Personnel Management: A New Approach*, London: Prentice Hall.

Torrington, D. and Hall, L. (1998) *Human Resource Management* (4th ed), London: Prentice Hall.

Torrington, D. Hall, L. and Taylor, S. (2005) *Human Resource Management* (6th ed), Harlow: Prentice Hall Financial Times.

Training Agency (1988) The definition of competences and performance criteria, guidance note 3 in Development of Assessable Standards for National Certification (p. 5), Series, Sheffield: Training Agency.

Trejo, D., Patil, S., Anderson, S. and Cervantes, E (2002) Framework for competency and capability assessment for resource allocation, *Journal of Management in Engineering*, Vol. 18, No. 1, pp. 44–49.

Tuckman, B.W. (1965) Development sequence in small groups, *Psychological Bulletin*, Vol. 63, pp. 384–99.

Turner, P. (2002) *Strategic Human Resource Forecasting*, London: CIPD.

Vakola, M. and Rezgui, Y. (2000) Organisational learning and innovation in the construction industry, *The Learning Organisation: An International Journal*, Vol. 7, No. 4, pp. 174–83.

van Eijnatten, F. M. (2004) Chaordic systems thinking: some suggestions for a complexity framework to inform a learning organization, *The Learning Organization*, Vol. 11, No. 6, pp. 430–49.

van Eijnatten, F. M. and Putnik, G. D. (2004) Chaos, complexity, learning, and the learning organization, *The Learning Organization*, Vol. 11, No. 6, pp. 418–29.

Volberda, H. W. (1997) Building flexible organizations for fast-moving markets, *Long Range Planning*, Vol. 30, No. 2, pp. 169–83.

Walker, D. H. T. (1996b) Characteristics of winning construction management teams, in Langford, D. A. and Retik, A., eds, *The Organisation and Management of Construction: Shaping Theory and Practice*, Vol. 1, pp. 321–34.

Walker, D. H. T. and Loosemore, M. (2003) Flexible problem solving in construction projects on the National Museum of Australia project, *Team Performance Management: An International Journal*, Vol. 9, No. 1, pp. 5–15.

Weddle, P. (1998) Career fitness, *Journal of Management in Engineering*, Vol. 14, No. 2, pp. 23–25.

Wickisier, E. (1997) The paradox of empowerment – a case study, *Empowerment in Organisations*, Vol. 5, No. 4, pp. 213–19.

Wilkinson, A. (2001) Empowerment, in Redman, T. and Wilkinson, A., eds, *Contemporary Human Resource Management*, Harlow: Pearson, pp. 336–52.

Williams, H., Tansley, C., *et al.* (2008) Skills and Knowledge of HR IS Project Teams: A Human Capital Analysis, in *The Second European Academic Workshop on e-Human Resource Management: 'e-HRM: Barrier or Trigger for HRM Transformation?'* Aix en Provence (Carry Le Rouet), France.

Wood, S. and de Menezes, L. (1998) High commitment management in the UK: evidence from the workplace industrial relations survey and employers' manpower and skills practices survey, *Human Relations*, Vol. 51, No. 4, pp. 485–15.

Wysocki, L. (1990). Implementation of Self-managed Teams within a Non-union Manufacturing Facility, Paper presented to the *International Conference on Self-managed Work Teams*, Denton, TX.

Yankov, L. and Kleiner, B. H. (2001) Human resource issues in the construction industry, *Management Research News*, Vol. 24, No. 3/4, pp. 101–5.

Index